Smartland Korea

perspectives on CONTEMPORARY KOREA

SERIES EDITORS: NOJIN KWAK AND YOUNGJU RYU

Perspectives on Contemporary Korea is devoted to scholarship that advances the understanding of critical issues in contemporary Korean society, culture, politics, and economy. The series is sponsored by The Nam Center for Korean Studies at the University of Michigan.

Hallyu 2.0: The Korean Wave in the Age of Social Media
  Sangjoon Lee and Abé Mark Nornes, editors

Smartland Korea: Mobile Communication, Culture, and Society
  Dal Yong Jin

Smartland Korea

Mobile Communication, Culture, and Society

Dal Yong Jin

UNIVERSITY OF MICHIGAN PRESS

Ann Arbor

Published in the United States of America by the
University of Michigan Press
Manufactured in the United States of America
♾ Printed on acid-free paper

2020   2019   2018   2017   4   3   2   1

A CIP catalog record for this book is available from the British Library.

Library of Congress Cataloging-in-Publication Data

Names: Jin, Dal Yong, 1964– author.
Title: Smartland Korea : mobile communication, culture, and society / Dal Yong Jin.
Description: Ann Arbor : University of Michigan Press, [2017] | Series: Perspectives on contemporary Korea | Includes bibliographical references and index.
Identifiers: LCCN 2016031596| ISBN 9780472053377 (pbk. : alk. paper) | ISBN 9780472073375 (hardcover : alk. paper) | ISBN 9780472122615 (e-book)
Subjects: LCSH: Information technology—Social aspects—Korea (South) | Smartphones—Social aspects—Korea (South) | Mobile communication systems—Korea (South) | Technological innovations—Social aspects—Korea (South)
Classification: LCC HN730.5.Z9 I5646   2017 | DDC 303.48/33095195—dc23
LC record available at https://lccn.loc.gov/2016031596

Contents

Preface

In the early 21st century, the phenomenal growth of smartphones has profoundly changed people's lives. The smartphones and relevant applications that people use create a new culture in which people communicate, connect, and entertain in new ways. The smartphone is indispensable for connecting with friends and relatives, not only through its telephony function but also through instant mobile messenger apps, such as Kakao Talk, Line, WeChat, and WhatsApp, and for enjoying all kinds of cultural activities. Mobile games on smartphones, far more advanced than those on feature phones, have also shifted the map of the digital game industry and global game market, as many players have migrated from console, handheld, or online games to mobile games. Smartphones also function as new capital resources for media and telecommunications corporations, as well as corporations that are built on digital platforms, such as Facebook and Twitter.

Because of the increasing significance of the smartphone and related technologies, many countries, both Western and non-Western, have developed their own smartphones and apps since the early 2000s. Korea in particular has become one of the major hubs for smartphone technologies, culture, and digital economy. Although Korea once lagged behind in the penetration of smartphones and their apps, it has unexpectedly become one of the powerhouses in the global information and communications technology (ICT) industry. As mobile technologies have become among the most important ICTs, smartphones have become the most innovative and cutting-edge technology in networked Korean society, adding to its already notable resume of digitally mediated modes of telecommunications. The invention of the iPhone entailed not just a handset, but an ecosystem, and the smartphone explosion is creating a ripple effect across Korean telecommunications industries and the general public. Smartphones have shaken the mobile communication system and altered people's daily activities. Korea thus exists as a test bed for the future of smartphone-related technology that policymakers, technology experts, media, and telecommunications scholars around the world keep an eye on.

The emergence of smartphone technologies and culture, including instant mobile messenger apps, in Korea can be explained by referring to supportive government policies, competition among telecommunications firms (both gadget and service corporations), and fervent IT consumers. Customers have played a primary role in the growth of the smartphone, app economy, and culture. Young Koreans' engagement with smartphones and related apps suggests that the smartphone has become a symbolic and material resource for young people's urban lifestyle.

This "smartland," in which smartphones and apps are not only created but are also essential features in people's daily activities, has led to sociocultural problems. In the smartphone era, significant social concerns, including smartphone addiction, digital bullying, and privacy invasion, have emerged. The smartphone generates a new form of digital divide, the smartphone divide, resulting in the intensification of social inequalities. While smartphones are no longer luxuries for many, some Koreans still suffer from both inherited and newly emerging socioeconomic problems due to the lack of information and available funds. The development of smartphones might be presumed to reduce existing social problems; however, new technology has always been part and parcel of human hierarchy and domination. The dominance of U.S.-based operating systems has increased, and U.S.-created platforms such as Facebook, Google, and Apple reap the benefits of the emerging app economy.

As Korea plays its part as a test bed in the realm of new media, and in particular the smartphone, it is worthwhile to analyze mobile communication, culture, and society in the context of Korean smartphones. This is particularly the case because the world at large knows little about how the local smart mediascape is articulated, or how local consumers, followed by global consumers, appropriate such technologies and create smartphone-related cultures. Of course, our focus should not be a celebratory account of the achievements of local smartphone culture and industry, nor a pessimistic deluge of jeremiads on the dark side of the era, but the significance of the Korean social milieu in the development and growth of smartphones. In other words, it is critical to locate the emergence of smartphones within the growth of mobile technologies and overall wireless telecommunications industries embedded in Korea's ICTs.

I therefore analyze the role of the smartphone as one of the major components of Korea's national economy and culture. From mega ICT corporations, such as Samsung and LG, to small start-ups, such as mobile game firms and app developers, local companies have invested in the development of smartphones and apps because these new technologies are a lucrative business. By examining the role of smartphones and apps, I aim to comprehend whether networked activities and information industries in conjunction with smartphones and apps function as significant parts of the app economy and youth culture. How smartphone uses have taken shape within the context of Korea's particular mobile culture is a key agenda, given that the users are the primary actors. In this regard, I explore some of the sociocultural factors contributing to the growth of mobile games and youth culture. I firmly believe that an analysis of the rise of Korea's smartphone technologies and culture requires a comprehensive exploration of the country's experiences beyond ICT.

This carefully documented study of Korea's experience of sociocultural and political economic change within the smartphone system will shed light on more general trends in the shifting global mobile communication system. It will ask readers to contemplate how key features of global mobile communication have been reorganized and restructured within the context of global political-economic shifts and accompanying technological breakthroughs.

Finally, chapter 8 in a previous form was presented as a coauthored paper with Kyong Yoon in *Convergence: The International Journal of Research into New Media Technologies* 22, no. 5 (2016): 510–23. I also want to acknowledge that chapter 6 was supported by a National Research Foundation of Korea Grant funded by the Korean government (NRF-2013S 1A3A2054849).

PART I

Innovation and Mobile Communication

1

The Rise of Smartland Korea

> For such a small country, South Korea sure is in the news a lot. Occasionally that news is grim. Sometimes it's just plain astonishing. From the weird to the wonderful to sci-fi stuff from a Samsung galaxy far, far away, here are things Korea pulls off more spectacularly than anywhere else. Want to see what the future looks like? Book a ticket to the country with a worldwide high 82.7% Internet penetration. Among 18 to 24 year olds, smartphone penetration is 97.7%. While they are chatting away on emoticon-ridden messenger apps such as Kakao Talk, Koreans also use their smartphones to pay at shops or watch TV (not YouTube but real-time channels) on the subway. (CNN 2013)

The dramatic advent of smartphones has fundamentally changed people's daily lives. While "the bite-size software programs people loaded onto their mobile phones or tapped into on the Web several years ago" seemed to be "silly games and pointless novelties" (MacMillan and Burrows 2009), smartphones and their applications (hereafter apps) have created new capital for information and communications technology (ICT) corporations and shifted the ways people communicate. There was a time when people used mobile phones only as a medium of communication. In the networked society in the early 21st century, however, the smartphone is essential for people to connect with friends and relatives and to enjoy entertainment and cultural activities (Rainie and Wellman 2012). Mobile communication through the smartphone is becoming embedded in society (Ling 2012, vii).

Increasing public adoption of smartphone technologies has especially expanded the digital economy—an economy with an increased emphasis on informational, global, and networked activities and information industries, such as computers, the Internet, and telecommunications (Castells 1997; D. Schiller 1999; 2012)—and people's daily culture. A smartphone functions as a handheld platform that has multiple features to run various applications. Smartphones have made it easier to download all kinds of apps, and the pervasive use of apps has enabled people to engage in mobile games while searching for information, from foods to maps (Goggin 2009). The increasing role of apps in the smartphone era creates a new form of digital economy, the app economy, a term that refers to the range of economic activity surrounding mobile applications, including the sale of apps, ad revenue within apps, and digital goods on which apps are designed to run (MacMillan and Burrows 2009; Jin 2014). At the same time, smartphones and relevant apps have fundamentally shifted people's cultural activities.

Many countries, both Western and non-Western, have developed their own smartphones, including Apple in the U.S., Nokia in Finland, Xiaomi in China, and HTC in Taiwan. These telecommunications and new media corporations have vigorously invested in the smartphone system in order to compete with each other in global markets. Among them, South Korea (hereafter Korea) has become a center for smartphone technologies and culture, as well as the app economy, with several leading domestic-based transnational corporations. Two smartphone makers, Samsung Electronics and LG Electronics, have rapidly grown to compete with Apple and Nokia, while Kakao Talk and Line—two free mobile instant messenger applications for smartphones—play a key role in the app-based economy and culture, like WhatsApp (developed in the U.S.) and WeChat (developed in China), both nationally and globally.

Although the country once lagged behind in the penetration of smartphones and apps—as the iPhone's debut in Korea was belated—it has become one of the powerhouses in global information and communication technologies. Korea was a comparative latecomer to the smartphone revolution, but it did not take much time for Korean smartphone makers to take over the market, surpassing Apple's iPhone in terms of the market share in the global smartphone markets (P. Kim 2011; D. Lee 2012; Jin 2014). As new technologies appear, Koreans are used to rapidly adopting them, faster indeed than any other nation. Korea is full of "early adopters" who are willing to buy newly released digital devices for consumer testing (Jin 2010).

The smartphone, as the most swiftly diffused technology in Korea in the second decade of this century, has replaced the feature phone and changed the form of social mediation. The shift from feature phones (with physical keys) to those of touch-sensitive smartphones has been precipitous. Domestically, Korea had 58.9 million mobile subscriptions as of December 31, 2015. The number of smartphone users among mobile users spiked to exceed 43.7 million by the end of December 2015, consisting of 74.1% of total mobile phone users, up from around 1.6% of total mobile phone users in December 2009 (Ministry of Science, ICT and Future Planning 2016) (fig. 1). Many other countries have also rapidly increased their smartphone penetration (Nielsen 2012a; ComScore 2014); however, Korea has evidenced one of the fastest—if not the fastest—adoptions of smartphones in the world. Consequently, Korea has also swiftly advanced its use of relevant apps.

It is significant that the penetration rate of the smartphone among Korean youth, in particular the age group from fourth graders to seniors in high school, as well as those in their twenties, is much higher than the national average, primarily because Koreans emphasize education and they give their children smartphones. The smartphone is a necessary tool for the youth in order to connect with not only their parents but also their private teachers. According to the Ministry of Gender, Equality and Family (2013), by the end of 2013 about 81% of Korean teenagers had smartphones—more than 10% higher than the national average. Since they are currently major users, and are also future customers, the smartphone will continue to grow in Korea as an important digital technology for the national economy and youth culture.

Globally, Korean-made smartphones, including Samsung's Galaxy series and LG's Optimus and G2, have competed with U.S.-based iPhones and increased their global market share since 2009. Back when Samsung still focused on feature phones in 2008 and 2009, its share in the global mobile phone market was 3.6% and 3.3%, respectively. In 2010, however, Samsung became the second largest mobile maker, with a 20.1% market share, and then became the number one maker and exporter, with 23.4% in 2012, surpassing Nokia (International Data Corporation 2012; 2013).

Equally important is the role of local smartphone applications, including Kakao Talk and Line—instant mobile messenger applications for free calls and messaging. The remarkable development of apps underlines that people are increasingly going online by means of smartphones and other wireless devices (Associated Press 2013), creating a new pattern of media convergence with smartphones. As smartphones have become a major part of people's daily lives, Korea has witnessed the rapid growth of apps and has exported the apps to several Asian countries and other parts of the

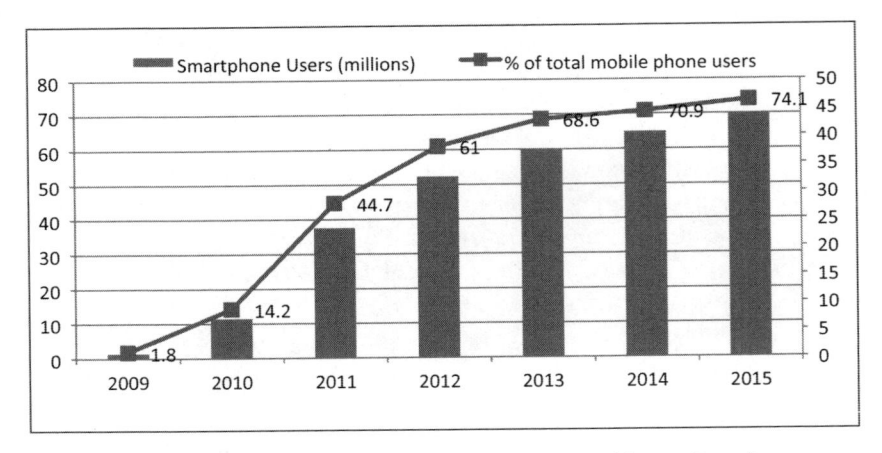

Fig. 1. The Growth of Smartphone Penetration in Korea. (*Source:* Data from Ministry of Science, ICT and Future Planning [Mirae Ch'angjo Kwahak Pu] 2014, 2015a, 2016.)

world. Kakao Talk and Line allow their customers to send and receive messages, videos, and photos for free.[1] Thanks to the convenience and simplicity of their message services and other functions, the number of subscribers has soared since their release in 2010 (Kakao Talk) and 2011 (Line) (Millward 2013).[2]

These two smartphone platforms have provided unprecedented functionalities. In addition to free calls and instant messages, Kakao Talk users are able to share various forms of content. Kakao Talk began as a free mobile instant messenger service, but has transformed itself into a platform for the distribution of diverse third-party apps and content, including mobile games, which users can play with their friends through the messaging platform. As Kakao Talk itself boasts, until a few years ago mobile game applications struggled to attract one million downloads. Kakao's game platform completely shattered this limit, and gave birth to eight games that had recorded more than 10 million downloads as of July 2011 (Russell 2013).

Consequently, Korea exists as an intriguing test bed for the future of smartphone-related technology, as Japan was with the development of mobile phones, known as *keitai*, early in the century (Ito et al. 2005; Daliot-Bul 2007).[3] As Sun-Sun Lim and Gerald Goggin (2014, 663) correctly observe,

Japan pioneered many of the technologies associated with cellular mobile devices. It made key contributions to first-generation mobile networks and handsets; in the 1990s, it was one of the first countries, where the Internet were to be notably encountered on mobile phones; mobile social software, the forerunner of today's mobile, social, and locative media, was pioneered, with iconic devices such as Lovegety; in the early 2000s, the camera phone was invented and first shipped with mobile handsets in Japan.

For its part, Korea was the test bed for various technical and social innovations, including mobile television, mobile games (as part of Korea's teeming online gambling culture), early social network systems such as Cyworld, camera phone culture, and locative media (Hjorth 2006; 2009; Hjorth and Chan 2009; Lim and Goggin 2014).

The recent emergence of the smartphone industry and culture as well as free instant messenger apps in Korea can be explained by a combination of favorable information technology policies, severe competition among information technology corporations, and enthusiastic IT consumers. Customers especially played a major role in the early development of the smartphone and app economy and culture in Korea. Young Koreans' engagement with smartphones and related apps suggests that the smartphone has become a symbolic and material resource for young people's urban lifestyles. For example, in the last few years, through converging smartphones and mobile gaming, the country has taken a big step toward Korea's goal of becoming a "mobile game wonderland" by appropriating smartphones as the platform of choice for mobile games, putting behind the days when the self-proclaimed tech giant found itself in a painful position as a smartphone laggard (K. S. Lee 2010). Accordingly, mobile technologies have become some of the most significant advances in the early 21st century, and smartphones have become the most innovative and cutting-edge technology in networked Korean society, adding to its already notable resource of digitally mediated types of communication.

"Smartland," in which leading smartphones and apps are not only created but also used in people's daily activities, is not without areas of concern. With the rapid growth of the smartphone, significant social problems, such as smartphone addiction, the digital divide, digital bullying, and privacy invasion, as well as foreign dominance in software, have emerged. The digital divide in the smartphone era assumes a new form, the smartphone divide, which causes social problems through its expansion of social inequalities. The roles of U.S.-based operating systems, namely Android and iOS, have also continued to grow; therefore, Google and Apple substantially benefit from the emerging app economy. New technology, including smartphones, may convert certain social issues into vestiges of the past; however, technology will always be a component of "the reality of human hierarchy and domination" (Demont-Heinrich 2008, 381). While smartphones are no longer luxuries for most Koreans, some still suffer from both persisting and newly emerging socioeconomic problems in the era of smartphone.

Major Goals of the Book

Despite the significance of smartphones in the digital economy (partially in the form of the app economy) and in youth culture across the globe, there has been a lack of academic literature exploring how smartphones are integrated into the socioeconomic and cultural landscape of a particular context, and how young smartphone users engage in it. Academic work on smartphones has been burgeoning in Western countries in recent years; however, there are only a few works on the evolution from mobile phones to smartphones in Korea (Ok 2011; D. Lee 2012; Lee et al. 2015), and this may be a reflection of the short history of the phenomenon. There has been no single book written in English about the emergence of Korea's smartphones in the local scene.

Of course, there are a few book-length publications (Oh and Larson 2011; K. S. Lee 2011; Mahlich and Pascha 2012; Hjorth et al. 2012) focusing on the recent growth of telecommunications systems. These books are valuable sources, and many scholars and students have learned about the scope of Korea's mobile technologies and cultures through them. Among these, Kwang Suk Lee (2011) wisely analyzed the rapid growth of information technology by focusing on broadband systems in tandem with government policies through the lens of political economy. The primary agenda of the book emphasized the nexus of the Korean government and ICT corporations in building national infrastructure in the pre-smartphone era. Myung Oh and James Larson (2011) also documented the growth of ICTs in the 1980s and the 1990s, and they discussed the proactive role of the Korean government, which propelled the nation's broadband revolution in the 1990s and its mobile phone development in the early 21st century. Meanwhile, Whasun Jho (2013) examined the uniqueness of Korean regulatory reforms of the mobile telecommunication sector and argued that market-oriented regulatory reforms and liberalization can be explained by focusing on the interactions among the state, the private sector, and international political economic movement. However, Jho did not analyze the entire scope of the Korean mobile communication system, including smartphone technologies and cultures.

In sum, these previous works have not discussed the major characteristics of smartphones and apps per se, partially because they did not have enough time to witness the recent growth of smartphone technologies and culture. A handful of chapters in edited volumes (Hjorth et al. 2012) have also considered limited cultural genres without discussing the sociopolitical context, including cultural policies and creative industries. The lack of empirical and theoretical studies seems to leads to a gap in the literature on smartphones—the overlooked aspect of local diversity.

In fact, while the smartphone mediascape has recently been examined from the perspective of the convergence of different media (Goggin 2010; Sinckars and Vonderau 2012), only a few contextualized empirical studies address the locality of smartphone culture (Hjorth et al. 2012). In recent smartphone studies, the local dimension is addressed at best as a site in which a global technology is disseminated and consumed, while the production side of a local smartphone mediascape is largely overlooked (Jin and Yoon 2016). Thus, despite the explosive influx of imageries of smart technologies into our daily lives, people know little about how the production side of the local smart mediascape is articulated with the ways in which local consumers appropriate such technologies. In this respect, we are required to empirically examine how smartphones are produced, appropriated, and thus redefined in their local contexts for a comprehensive understanding of what smartphones mean to youth culture and the digital economy.

This study, as the first book-length English-language treatment of smartphone communication in Korea, attempts to comprehensively analyze mobile communication using smartphones. As Mark McLelland (2007) points out, there is no reliable overview that attempts to theorize the regional growth of mobile technologies, now, of course smartphone technologies, and influences such as the immense impact that Korean technologies have had on the global market. Therefore, this book investigates focuses of inquiry neglected until now—a localized communications landscape emerging with the smartphone and its apps. I aim to make a significant contribution to the literature, primarily by means of the book's comprehensive analysis of the emergence of the Korean smartphone mediascape. As Raymond Williams (2003) points out with respect to television, technology is complex and ultimately transcends its elements, its operations, uses, and contents, and is rather a cultural form constituted by shared knowledge and social practices. That is how I analyze the smartphone in Korea.

My focus is not the celebratory achievement of local smartphones, nor a pessimistic glimpse at the dark side of the era, but the significance of social milieu in the development of smartphones. I situate the emergence of smartphones within the growth of mobile technologies and overall telecommunications industries embedded in Korea's information and communication technologies. Therefore, I historicize the growth of the current form of smartphone technologies and culture not as separate but as continuous developments in tandem with telecommunications systems. I also document the role of the Korean government under neoliberal globalization, given that changing ICT policies in conjunction with economic considerations of the country have played a pivotal role. Several theoreticians (Ohmae 1995; Appadurai 1996; Giddens 1999; Tomlinson 2000; Hardt and Negri 2000) have argued that globalization diminishes the relevance of borders and territory, and thereby undermines the territorial nation-state's role as the major institution for governing both national and global affairs. This book therefore analyzes whether the nation-state in conjunction with smartphone businesses has lost its function or has maintained or intensified its role.

Second, this book discusses the smartphone as one of the major elements of Korea's app economy. As I have already discussed, with the rapid growth of smartphones and applications, many countries, both Western, including the U.S. and the U.K., and non-Western, including China and Korea, have developed smartphones and applications to advance sociocultural, technical, and economic growth. In Korea, local companies, from mega ICT corporations, such as Samsung and LG, to small start-ups have invested in smartphones and apps because they have become a lucrative business ever since Apple launched the Apple Store based on the huge success of the iPhone in 2009 (MacMillan and Burrows 2009). In particular, the app economy—the range of economic activity surrounding mobile applications—depends not only on infrastructure but also on software (Jin 2010; 2014), and the evolution of smartphones has consequently influenced the software sector in Korea. By analyzing the role of smartphones and apps at the same time, I aim to understand whether networked activities and information industries in conjunction with smartphones and apps work as significant parts of the national economy.

Third, I map out important dimensions within globalization theory. The rapid growth of the smartphone has raised significant legal debates.

Since Apple released the iPhone in January 2007, mobile phone makers around the world, including Samsung and LG in Korea, and HTC in Taiwan, have developed their own smartphones. These major smartphone makers have argued that other corporations copy their unique intellectual property (IP), including patents and design, and, as key forms of IP, patents and design have emerged as the new scene in which competition between manufacturers of smartphones has played out. Therefore, the growth of Korea's smartphones cannot be examined without reference to their close relationship with global forces, including both Western countries and Western-based transnational corporations, including Apple.

It is also significant that platforms dominated by U.S. Android phones seem to be everywhere, and they have developed a hegemonic power in the smartphone industry, just as Windows and Mac have in personal computers. Although Samsung has become a global leader with its gadgets, as far as software is concerned, the U.S. is still the major player. This is why it is critical to understand the overall smartphone system, including operating systems, which allows us to better understanding the global trend in the realm of smartphones.

Many policymakers and media chant about the rapid growth of Korea's smartphones in the global markets, and they argue that the local force has become a major international power. However, when our discussions reach the subjects of operating systems, the U.S. has been consistently dominant, and has even intensified its hegemonic power. I thus critically examine significant conflicts between global and local forces. I especially investigate the power struggles between the global (iPhone) and the local (Galaxy) products in order to determine whether the emergence of local smartphones has changed the contour of globalization theory.

Finally, I analyze how smartphones have taken shape within the context of Korea's particular mobile culture, given that users are the primary actors in the era of smartphones. I explore some of the sociocultural factors contributing to the growth of smartphones and relevant apps in Korea. I then discuss the ways in which the emergence of smartphone use has shaped the development of Korea's youth culture, compared with what one might find in other youth cultures in their own national context. This part of the analysis includes two major youth subcultures in Korea: mobile gaming and the use of Kakao Talk. As Gerald Goggin (2007, 133–34) has pointed out with regard to pre-smartphone mobile technologies, "The field of mobile studies is a fascinating interdisciplinary one; however, there has not yet been the sustained concentration on questions of culture that technology, its meanings, practices, and consumption merit. At this conjuncture, attention to the cultural aspects of mobiles is something sorely needed."

Korea is one of the largest markets in online gaming; therefore, it is not surprising to witness the growth of smartphones and mobile gaming after the Korean government lifted once unfavorable rules that restricted game content in smartphones, because Korean youth have been regarded as enthusiastic early technological adopters typically willing to buy newly released digital devices for consumer testing (Yoon 2006; Jin 2010). Mobile gaming has become one of the most exciting youth cultures and lucrative game businesses in very recent years. The nascent emergence of the smartphone industry and culture in Korea cannot be understood without discussing the role of local smartphone apps that have provided convenient platforms for local ICT users. While there are several important apps, Kakao Talk especially offers a vivid example of how smartphone evolution is reimagined in a local context. Young Koreans' engagement in Kakao Talk and related mobile apps suggests that smartphones have become symbolic and material resources for their urban lifestyles.

Overall, through an examination of Korean smartphone communication in light of its sociocultural elements, converging political economy, and cultural studies contingencies, I investigate some of the underexamined complexities inherent in the conception, development, implementation, and reception of smartphones in a global arena. An analysis of the rise of Korea's smartphone technologies and culture depends on a comprehensive exploration of the country's experiences with the expansion of digital technology. Detailed accounts of the revolutionary developments of smartphone technologies and culture in the second decade of the century provide a useful case study, because a carefully documented account of Korea's experience of sociocultural, political, and economic change within the smartphone system sheds light on more general trends of the shifting global mobile system. In other words, this study will lead us to think about how some of the key features of the global mobile system have been reorganized and transnationalized since the early 1990s, that is, how the transformation of the global mobile system can be understood within the larger context of global political-economic shifts and accompanying technological development.

Theoretical and Methodological Frameworks

I employ a few theoretical frameworks; in particular, I utilize political economy perspectives in terms of sociohistorical and institutional analyses and combine them with cultural analyses, interpreting resistance in light of the power imbalance intensified by media convergence and corporate synergy on the structural level. As Dan Schiller (2014, 6) aptly puts it, "The role of information and communications needs to be sought within the political economy's chief developmental processes," including capital's reorganization of the production process. As the book's major goals are to identify both the primary reasons for the growth of the local smartphone industries and document the continuity and change of the smartphone sector, which fundamentally influences people's lives, it is crucial to understand the increasing role of smartphones and apps within a bigger socioeconomic and cultural milieu. Indeed, ICTs, in this case, smartphones, play an integral role in economic development and innovation

What we also have to keep in mind is that sociocultural developments influenced by the advancement of ICTs cause imbalances and inequalities both within a country and across borders in the global political economy. Evidenced in several previous ICTs, including the Internet and mobile phones, as well as video games, the global production, distribution, and consumption of these new technologies have been the result of power struggles between a few dominant countries and transnational corporations in these developed countries and the remaining countries. In this respect, Christian Fuchs (2014, 17) points out the significant role of the political economy by saying, "Critical theory can help to explain the causes, conditions, potentials and limits of struggles. Critical theory rejects the argument that academia and science should and can be value-free. It rather argues that political worldviews shape all thought and theories." As Vincent Mosco (2014, 1–2) clearly explains with the case of cloud computing, "The global expansion of ICTs controlled by a handful of companies continues a process of building a global information economy. ICT corporations that once housed an information-technology department with its craft tradition can now move most of its work to smartphones and apps, where IT functions and its labor are centralized in an industrial mode of production and distribution." Smartphones and apps also take the next step in a long process of creating a global culture of play with mobile games. Therefore, it is critical to understand the power relationships between the political and the economic in the global flow of smartphones and apps.

However, it is also essential to enter into a discussion of the emerging smartphone sector by utilizing diverse perspectives, given that technology is no longer an isolated area of study. As Arnold Pacey (1985, 4–6) argued, those who write about the social relations and social control of technology should know the cultural aspects as well as the organizational and technological aspects. Toby Miller (2006, 6) also points out that "every cultural and communications technology has specificities of production, text, distribution and reception." These sentiments reflect my rationale in investigating the smartphone sphere in ways that reflect its position as a sector that no longer exists in a vacuum, distinct from others. As Lev Manovich (2013, 7) states, "Platforms, such as Google, Facebook, iOS, and Android, are in the center of the global economy, culture, social life, and, increasingly, politics"; therefore, it is critical to analyze the growth of smartphones and apps and their impacts through a nuanced and holistic approach.

As several theoreticians (Richardson 2012; Chan 2008; Hjorth 2012) claim, the heuristic approach to new technologies is indeed essential. Among these theorists, Richardson (2012, 135) argues that "the heuristic understanding of the nexus of the cultural, organizational, and technological aspects is necessary in order to map out the emergence of new technologies." She (2011, 419) points out that "mobile and handheld gaming presents a complex interplay of cultural, contextual and corporeal factors. The mobile phone and handheld game device are simultaneously— and often equally—acoustic, visual and haptic mediums." Hjorth (2012, 194–95) also makes an insightful interpretation of smartphone culture, including mobile gaming, deriving from social constructivism theory.

I am confident that a heuristic understanding of the growth of smartphone technology and culture can be achieved through the convergence of political economy and culture. While scholarship on the political economy of media "takes it as axiomatic that the media must be studied in relation to their place within the broader economic and social context" (Winseck 2011, 4), cultural studies emphasizes sociocultural processes in the growth of smartphones, and in particular, the reasons why people adopted the smartphone within such a short period of time. By employing a multiplicity of frameworks, therefore, I emphasize that the era of smartphones must be defined based on the sociocultural specificity of Korean smartphone use.

Through this sociotechnical examination of transformative mobile technologies and culture, from feature phones to smartphones, I hope to illuminate some of the complexities inherent in examining mobile platforms as they continue to be manifested in Korea. I pay attention to the change and continuity in Korea's smartphone landscape, and I chronicle political, economic, technological, and cultural dimensions that have led to this dramatic change in the global mobile system.[4] These theoretical frameworks converge with social constructivism theory to help us understand the rapidly shifting Korean mobile system, and readers will gain a more concrete grasp of the transformation of global mobile technologies and culture.

As for methodological considerations, several scholarly analyses of and discourses about mobile technologies and culture previously focused on local popular culture, employing audience studies to consider identities among youth and changing lifestyles in the development of the Korean mobile phenomenon. Others emphasized the role of the mobile industry as significant for the national economy. However, mobile culture is no longer a discrete and distinct sector. All its circuits, technology, texts, and promotion have become intertwined with the wider orbits of digital networks as critical zones for growth and profits (Klein et al. 2003, 176; Jin 2010; 2015). Cultural products should be defined based on specific combinations of technical, social, cultural, and economic characteristics and not exclusively on any one of these alone. Accordingly, I employ political economy approaches, emphasizing historical documentation of the growth of smartphones, and combine them with in-depth interviews gathered in fieldwork to strengthen the major analytical framework. As a body of empirical data, these interviews and observations provide an opportunity for those who live in smartland to express their ideas about topics that tend to be neglected in everyday conversations. Indeed, the insights gleaned from them will be explored in this book.

More specifically, I conducted interviews with a set of 50 young mobile phone users (19–29 years of age) who mostly used smartphones, not feature phones, in both Korea and North America between summer 2012 and winter 2014. The interviews covered their use of smartphones and engagement in mobile games, and the broader life circumstances. Thus, we can examine the roles played by these technologies in their lives. Participants were recruited primarily by means of referral within university networks in both regions. In Korea, 16 were office workers at information technology firms, including Samsung and LG. In North America, the interviewees were mostly familiar with Korean digital economy and culture, including smartphones and free communication apps. Interviews were conducted for two hours in a semistructured way that enabled interviewees to express their experiences and opinions freely. Candidates were selected for interview only if they had experience using a smartphone and playing mobile games, as well as using other apps.

I also used several documents, including government data, corporate reports, such as annual reports and financial statements, and publications by a few international agencies, including the International Telecommunication Union (ITU) as supporting materials. These governmental and corporate documents are excellent sources of information in that they provide detailed data on industry and corporate activities, as well as key policy issues.

Organization of the Study

The study is organized as follows. In chapter 2, I historicize the evolution of the smartphone. The narrative clarifies the multiple causes that led to the emergence of the smartphone and its distinct characteristics, including convergence and participatory culture as well as telecommunications policy, providing core sociohistorical foundations on which future debates may be based. It discusses the major reasons for the rise and fall of the first smartphone, shedding light on the sociocultural milieu in which the smartphone originated. Finally, the chapter examines the role of transnational capital in the development of smartphone technologies, focusing on whether it played a key role in the development of the smartphone over the last 20 years.

Chapter 3 documents the recent growth of smartphones and relevant apps as the transition toward another juncture of the Korean new media scene. This chapter discusses the rapid development in Korea of several ICTs, including broadband and mobile technologies, based on its unique sociocultural and technological milieu, including Korean's propensity to adopt new technologies quickly. The chapter documents the evolution of mobile phones and examines the major characteristics of the technology and industries in the bigger social and cultural context. It analyzes Korea's smartphone technologies and policy issues as part of the continuous development of the local telecommunications system. It investigates the ways in which Korean mobile telecommunications industries have been altered by domestic political-economic factors and by the current transformation of the global telecommunications industry.

Chapter 4 addresses the theoretical question of globalization in the era of smartphones by analyzing Korea's reception of the iPhone in 2009 and the recent growth of Samsung's Galaxy in the global market. It attempts to develop new perspectives on the pertinence of globalization to smartphones. It focuses on the new material conditions of globalization, which is constructed by the worldwide electronic network of capital. Globalization, which is based on mobility and connectivity, signifies a power shift of capital, and forces local states to affiliate with a part of the new world system. The survival of local regions depends largely on their close links to the global electronic conduits of capital (K. S. Lee 2008). This chapter situates Korea's smartphone growth within the universal structure of the electronic empire, which strives to enlist the local as an active part of the new global network.

Chapter 5 explores key aspects of smartphones and apps services in Korea and their implications for the app economy, given that Korea has become the world's best laboratory for smartphones and apps—and a place to look to for answers on how the app economy may evolve. It analyzes the rapid growth of smartphones and apps in the socioeconomic environment specific to the country. It recognizes technology as a socioeconomic product that has historically been constituted by certain forms of knowledge and social practice; therefore, this chapter examines the role of people in the diffusion of app services, and the implications of their role for the networked society.

In chapter 6, I critically attempt to develop new perspectives on the digital divide by discussing its pertinence to smartphone technologies. I contextualize the digital divide and its implications using less linear explanations of the divide than statistical measurements. The chapter examines the ways in which the digital divide can be understood in the smartphone era and analyzes its implications for our smartphone-driven networked society. It discusses major policy implications, including social inclusion. In addition to a historical approach contextualizing the recent growth of ICTs and relevant sociocultural issues, I rely on survey research that was conducted in December 2014. This hybridized methodological framework moves research and inquiry forward in ICT studies, analyzing current sociocultural factors behind the rapid growth of ICTs in Korea.

Chapter 7 analyzes how mobile gaming has taken shape within Korea's particular online gaming culture, the conditions of which have favored the PC platform. I explore some of the sociocultural factors contributing to the growth of mobile gaming on smartphone platforms. I then discuss the ways in which the emergence of smartphone use has shaped the development of Korea's mobile games, compared with what one might find in other cultures and national contexts. Finally, I articulate how users perceive their own changing conditions amid the process of the rapid diffusion and growth of mobile games. In particular, I explore the transfer between online gaming culture and mobile gaming culture in a specific subset of Korean youth in a highly networked, urban setting.

In chapter 8, I examine a localized media environment arising with the smartphone and apps, with particular reference to young Koreans' engagement in a local app platform, Kakao Talk. I analyze what I call the "Kakao talkscape" as a form of mediascape embedded in and constructed by specific sociocultural circumstances in Korea. I attend to the ways in which users have engaged with the rapid socioeconomic progress of smartphones and Kakao Talk. I have employed a qualitative methodology consisting of in-depth interviews with young smartphone users, along with a sociohistorical analysis. The participants were asked to talk about their use of smartphones, engagement in Kakao Talk, and the broader life circumstances in which one can examine the roles played by these technologies in their lives.

Chapter 9 summarizes the major characteristics of smartphone technologies and culture embedded in Korea, and discusses whether we need to develop non-Western media theories to explain the rapid growth of local smartphone technologies and culture in the global markets or whether current forms of these theories are applicable. This final chapter also discusses what we have to keep in mind in further studies on mobile technologies and cultural discourse in the currents of globalization, with Korea serving as a good case study for several emerging local markets.

2

Evolution of Smartphone Technologies

Putting Smarts into the Phone

Smartphones have had a tremendous impact on society, and the touch-screen iPhone has revolutionized the world of communications. The image of the iPhone—sleek and urban in style—represents not only a uniquely modern, technological culture, but also a must-have, cutting-edge gadget in our digital platform-driven information society, in part because of its multifunctionality. The touch-screen smartphone is a life-changing technology for many: an all-in-one, portable device that combines the functions of a cellular phone with the functions of a computer. The iPhone led to the restructuring of the communication system by modernizing its structure and changing youth culture. "Rarely have technical innovations changed everyday life as fast and profoundly" as the smartphone, even if we acknowledge the significant influences of previous technologies, such as the typewriter, telegraph, and the Internet. Consequently, over the last several years, smartphone services have grown "from being niche market applications to globally available components of daily life" (Fehske et al. 2011, 55).

While the iPhone was the first smartphone to get the user interface, the iPhone was not the first smartphone. The iPhone has propelled smartphone technologies to new heights; however, it does not represent the original smartphone platform. The hundreds of millions of people currently using smartphones often take these devices for granted. They seem to forget the environment in which smartphone systems has evolved.

There are no academic discussions devoted to the evolution of the smartphone, although many works analyze the impact of iPhones and smartphones in general on our digital economy and culture. Although there are several compelling academic works on the origin of the mobile phone (Agar 2003; Dunnewijk and Hulten 2005; Engel 2008; Klemens 2010), which dates back to the 1940s, the relatively modern history of the smartphone has not been written yet.

This chapter historicizes the evolution of the smartphone. There are several significant predecessors, in particular, IBM's Simon, which is known as the prototype of the smartphone. A historical approach is very useful in understanding the causes behind the changing process of media and technology. The smartphone era may be divided into three major periods, including the prelude (up to the early 1990s), the first smartphone (IBM Simon) (1992–97), and the diversification of the smartphone, the decade from 1997 until 2007, when Apple launched its iPhone. I formulate these three time periods to clarify the relationship between socioeconomic transitions and the accompanying changes in the mobile industry.[1]

Employing the social construction of technology as its theoretical framework, this chapter clarifies multiple causes that led to the emergence of the smartphone and elaborates its distinctive characteristics of the smartphone, such as media convergence, participatory culture, and Korean telecommunications policy, providing core sociohistorical foundations upon which scholarly debates can be conducted. As Brian Winston (1998, 2) points out, "The concept of the Information Revolution is implicitly historical," and we cannot know that "a situation has changed—has revolved—without knowing its previous state or position. Even the notion of a Digital Age implicitly posits other preceding non-digital ages." This chapter discusses reasons for the rise and fall of the first commercial smartphone, shedding light on the sociocultural milieu in which the smartphone originated. Finally, it examines the role of transnational capital in the development of smartphone technologies.

How to Understand the Invention and Diffusion of Technologies

The social construction of technology (SCOT) emerged as an alternative to technological determinism, referring to the belief that technology advances along a path of its own following. "An invention, once introduced into society," inevitably and irreversibly taking an active predicate and progress along the way (Marx and Smith 1994, xi; Campbell and Russo 2003). Technological determinism explains that new technologies are the primary cause of major social and historical changes at the macro level in the social order, as well as being microlevel influences in how people view and use technological devices (Chandler 1996). Technological determinism was criticized mainly because it implied that "people were not accountable for technologies they used, because the path of technological evolution was viewed as one that was followed, not created" (Campbell and Russo 2003, 317). Social constructivism argues that people shape technologies, rather than technologies shaping society (Winner 1977; Mackenzie and Wajcman 1999).

Social constructivism implies that technological change cannot occur because new tools and processes prove clear-cut supremacy over other ways of doing things. Instead, the analysis needs to explain "why certain technologies are assumed to work better than others" (Mackenzie and Wajcman 1999; see also Volti 2008, 39).[2] To explain why things turned out the way they did, "Social constructivists describe how social structures and processes have affected choices of technologies" (Volti 2008, 39). According to social constructivist perspectives (Bijker et al. 2012; Flanigan et al. 2010), technological design is particularly a function of interconnected social, cultural, economic, and technical elements. Technological artifacts thus result from a complex interaction among several individuals, groups, and organizations. In fact, as Wiebe Bijker et al. (2012, 40) argue, "technological artifacts are culturally constructed and interpreted." As a result, technologies reflect felt social needs and current technical capabilities (Mackenzie and Wajcman 1999) and are both the result of and the impetus for social behaviors. Since the presence of interest groups and unequal distributions of power are fundamental aspects of every society, "Social constructivists are particularly interested in delineating the main actors involved in the development and selection of particular technologies, and in noting how their actions reflect their positions in society" (Volti 2008, 39).

Most importantly, as Fischer (1992, 16) points out, "Struggles and negotiations among interested parties shape the development of an innovation." "Given their focus on the complex interactions among diverse interest groups and technical capabilities, such [social constructivist] perspectives are useful in making sense of the origin and evolution of a wide range of technological artifacts" (Flanagin et al. 2010, 180). In other words, as Thomas Hughes (1994, cited in Volti 2008) claims, the strength of the social constructivist approach is its applicability in studying an early stage of development. Social, political, and economic forces are likely to exert the greatest influence when several alternative technologies emerge at about the same time.

Accordingly, the social constructivist approach is useful in investigating smartphone technologies, whose development, adoption, and use have grown exponentially since the 1990s, and exploded in the 21st century. Starting in the early 1990s, many innovators and corporations created smartphones. However, relevant social actors, "who are engaged in a process of defining technical problems, seeking solutions, and having their solutions adopted as authoritative within prevailing patterns of social use," cannot have equal power or opportunities; therefore, it is critical to comprehend who takes "potentially important choices that never surface as matters for debate and choice" (Winner 1993, 369).

However, I do not simply adopt what social constructivists argue regarding the role of relevant social actors, but critically address the different abilities of various actors who influence the outcomes of development and adoption, as Winner (1993) proposes. While scholars should admit the importance of the primary role of major players, it is significant to notice which decisions never land on the agenda, as a result of the imbalanced power relations among major actors or the elimination of potential actors. By noticing which issues are seldom articulated or legitimized, observing which groups are consistently excluded from power, "one begins to understand the enduring social structures upon which more obvious kinds of political behavior rest" (Winner 1993, 369). What we see and experience is the result of the successful actualization of ideas and technologies; therefore, it is necessary to analyze the invention and growth of innovative technologies through the social construction of technology.

The Origin of the Smartphone in the 1980s: When Did People Call It a Smartphone?

The iPhone and other smartphones have served people for only a few years; however, they have rapidly replaced the old form of mobile technologies, known as feature phones. The terms *cell phone* and *mobile phone* refer to a voice-centric cellular wireless device that has become the essential personal communication tool everywhere. In contrast, in the 21st century, a smartphone usually provides PIM (personal information manager) applications and some wireless communication capability in addition to traditional voice communication and messaging functionality (Zheng and Li 2006, 4). Smartphones are types of hardware architecture and software framework, converging several services, in particular computing and communication. In this sense, smartphones are about socializing and information gathering.

More specifically, in our contemporary society, the smartphone is a device sporting a large, sensitive screen instead of a regular keyboard (BellSouth-IBM Simon 1994)—although Blackberry was successful with a dedicated keyboard in the early iPhone era, at least for a while—having Internet communication features, complete personal digital assistant (PDA) functions, and, of course, all of the usual cell phone functions. "A smartphone is like a small, networked computer in the form of a phone." In other words, it is a cell phone that includes several software functions, such as e-mail and a browser. The very first generation of cell phones, despite their large size, could do little more than make phone calls. Later on, because of astounding advances in semiconductor technology, cell phones were equipped with more powerful processors, more storage, and a liquid crystal display (LCD) screen that made it possible to perform some computing tasks locally (Zheng and Ni 2006, 5).

The first smartphones did not include all of these features, which manufacturers added one after another. Therefore, it is hard to trace the starting point of the use of the term *smartphone* to characterize a device with one or more of the features I have mentioned. As Zheng and Ni (2006, 4) point out, "The term *smart phone* was initially coined by unknown marketing strategists to refer to a then-new class of cell phones that could facilitate data access and processing with significant computing power."

Arguably, though, the conceptualization of a smartphone materialized as early as 1973, when Theodore George Paraskevakos patented the notions of visual presentation within telephones. Paraskevakos filed paperwork with the US Patent Office in 1972 for "an apparatus for generating and transmitting digital information," and his company was ultimately granted the patent in May 1974 (Behrooz 2012). With the growth of digital technologies developed by Paraskevakos and others, the concept of the smartphone primarily started with the home telephone, not the cellular phone, between the mid-1970s and the early 1980s. Several major telecommunications corporations began to produce smart phones. Among these, American Bell introduced two smart telephones in 1983. The Genesis telesystem featured add-in cartridges and add-on modules that enabled the user to customize and reconfigure the Genesis terminal for each application. Initially, three cartridges were offered that provided enhanced voice capabilities—call waiting, call forwarding, and continuous redial of busy numbers (*Computerworld* 1983). Genesis telesystems were marketed through Bell Phone Centers and other retail stores. The other new American Bell terminal was called Touch-A-Matic 1600, which had a smaller screen and a more limited keyboard than Genesis (*Computerworld* 1983).[3]

At the same time, MCI Communications Corp, a long-distance telephone service company, planned to offer smart phones. "Since the difference between such devices and the previous group of prototypes is the operation of a widespread transformation (social necessity), it is likely, and history reveals common, that such creations occur in a number of places synchronously" (Winston 1998, 9). An MCI subscriber had to dial a local phone number and then enter a 5-digit code before placing a long-distance call. That put MCI at a disadvantage when competing against Bell Telephone's Long Lines, where those extra digits were not necessary. The phones MCI wanted to make were smart phones, with the ability to store and almost instantly dial any number stored in memory (Schrage 1983). Of course, these devices were only the beginning. Several manufacturers planned to make the phones even smarter by combining voice, data, and computing capabilities into one versatile whole.

The development of the concept (and later reality) of the smartphone can be seen in two more areas. First, the cultural milieu surrounding the use of the term *smartphone* provides some indications of the direction the technology was taking. It is interesting to know that the first references juxtaposed *smart* and *phone* or *smart* and *telephone*, instead of combining them in one word, *smartphone*. The emphasis was on smartness, not the phone, a reflection of the significance of software or digital functionalities. This is understandable because phones were already familiar, and were nothing new; but there had previously been no "smarts," which were new things. This kind of media attitude continued until the mid-1990s. As discussed in the following section, IBM Simon is commonly regarded as the first smartphone; however, people did not call it a smartphone, but "a smart cellular phone," and they emphasized again the term *smart(s)*, as the title of an article in *Advertising Age* ("BellSouth Puts Smarts in Simon Cellular Phone") in February 1994 exemplifies (Johnson and Fitzgerald 1994, 8). *USA Today* even called IBM Simon a "Super-phone" (or a superpowered cellular phone) in November 1993 (Maney 1993). The term *smartphone* (one word) appeared in the early 1990s.

Second, the development of the smartphone was also closely related to the changing social environment, as Bijker et al. (2012) aptly note. In the early 1980s, the automated office became more of a reality, and a wider array of hardware was competing for desk space. In an effort to keep room free for office workers, communications and computer manufacturers were combining telephone voice and data links in single desktop units, some with data-processing capabilities (Hozy 1985). While telecommunications companies produced smartphones to make money, they eventually made life a little easier for their users as well. The smartphone technically opened unanticipated avenues, and in the mid-1980s people in many countries, including the United States, the UK, and Canada began to use smart phones in both homes and offices.

In the early 1990s, the notion of the smartphone was further actualized in diverse areas, including airports, and many technicians and business reports predicted that telephones and telephone service would change more in the next few years than they had in the prior 20 years. In October 1991 Bill Husted (1991, 2), staff writer for the *Atlanta Journal and Constitution*, said, "The partnership of the telephone and the computer has created the first affordable smartphones. In the next few years, you'll be able to use your telephone to send printed electronic mail, check weather forecasts and pay your bill." One example of the marriage of computer and telephone was installed in two New York airports: John F. Kennedy International and Newark International, called "Public Phone 2000." The new phones had a built-in color screen and a computer keyboard (Husted 1991). This means that business travelers were able to send printed memos and messages from the keypad anywhere.

The Prelude: The Failure of AT&T's Smartphone in the Mobile Sector

The cellular concept emerged at Bell Labs in the late 1940s (Frenkiel 2010). Although mobile telephones had been around since 1946, it was not until the 1980s that the quality of frequency modulated sound, combined with reasonably priced microprocessors, digital switching, and a final decision on the cellular system spectrum combined to make it feasible to offer the first commercial cellular phone services in the United States (Agar 2003, 33–38; Federal Communications Commission 2005; Klemens 2010). Motorola filed for cellular patents as early as 1973,[4] and the FCC started auctions of cellular licenses on a city-by-city basis before the breakup of the Bell system in 1984 (Agar 2003; Dunnewijk and Hulten 2007). In early 1984, Motorola "introduced the first hand-portable cellular phone, the Motorola 9000, although it was hardly an instant commercial success" (Agar 2003, 42–43).

While there are several significant elements contributing to the growth of the first type of smartphone in the mobile telephone sector—innovative inventors, the development of digital technologies, and the increasing role of information and communications technologies corporations, as social constructivists argue (Volti 2008; Bijker et al. 2012)—changing government policies also influenced the growth of the concept of the smartphone. The story is closely linked to the growth of AT&T, followed by its breakup in the early 1980s.

AT&T, as the single largest telephone corporation until the early 1980s, planned to develop smart cellular phones. AT&T committed its vast resources to the creation of a nationwide cellular network in the late 1970s. In anticipation of FCC approval (and perhaps to encourage that approval) AT&T launched a full-scale development project, estimating that cellular service could be offered within about five years. The development went smoothly, but this prediction proved overly optimistic, as intense controversies over the role to be played by the AT&T monopoly delayed commercial service until 1983 (Sinclair and Brown 1983, A1).

In fact, in the U.S. after the launch of advanced mobile phone services (AMPS) in 1978, which was an analogue system, the first American cellular phone system came into operation in 1979 as a trial and went into commercial operation in 1983 (Young 1979; AT&T Archives 2011). "The first trial in America of a complete, working cellular system was held in Chicago in the late 1970s. The test began with a six-month trial from July 1978 to December 1978 involving 90 Illinois Bell employees as users, and expanded to include customers on December 20, 1978. . . . [W]hile AT&T had done most of the cellular phone development, government restrictions would keep it out of the emerging industry" (Klemens 2010, 50). At that time there were many small independent telephone companies, but "the AT&T monopoly (often called the Bell System) provided more than 80% of the local and long-distance telephone service in the U.S. Moreover, most of the equipment for the nationwide telephone network was designed by Bell Labs and built by Western Electric (AT&T's manufacturing arm)" (Frenkiel 2010, 14).

AT&T's dream could not be achieved primarily due to the breakup of the company into several smaller corporations. Two major changes in regulation in the early 1980s especially shifted the contour of the growth of the smartphone. The changes stemmed partly from the reorganization of AT&T by the FCC in 1981. First, until a decade before the breakup, AT&T had no competition, operating a government-endorsed monopoly in both the telephone equipment and telephone service businesses. In this period, virtually all phones were leased from the local phone company. AT&T owned the long-distance lines and local Bell operating companies provided local service and telephone equipment. But under the restrictions starting in the early 1980s, the local companies were limited to providing service: they no longer would be responsible for providing new equipment beyond what was in their inventory on December 31, 1982 (Sinclair and Brown 1981, A1). Equipment would come from the new, unregulated subsidiary of AT&T called American Bell and its competitors. American Bell intended to sell rather than lease telephone equipment to residential and single-line business customers, and the company needed a new telephone gadget to attract customers (Sinclair and Brown 1981).

This regulation eventually helped the development of the smartphone because, under this changing environment, local phone companies abandoned their traditional business of installing phones in customers' homes and businesses and charging a monthly rental for their use. The new competition would lower the cost of standard phones and accelerate the availability of a new generation of "smart" phones employing computer technology—cordless phones and automatic dialing features among them (Sinclair and Brown 1983, A1). Many smartphone users these days may not see cordless phones as smart enough; however, these shifts were significant enough to be recognized as resulting in "smart" phones, differentiating them from traditional cord telephones (Schrage 1983).

Second, the smartphone picture became more complicated for AT&T on January 1, 1984, when AT&T was forced to divest itself of local Bell companies under an agreement signed with the Justice Department ending a long antitrust suit. The 22 former Bell operating companies became separate and independent regional regulated telephone monopolies. AT&T remained a vertically integrated company consisting of Long Lines, Western Electric, and Bell Labs, but it was no longer in the business of providing basic local switched telephone services (Horwitz 1986; Klemens 2010). However, the antitrust division of the Department of Justice, which had sought to limit AT&T's monopoly power for most of the century, considered breaking up AT&T itself. "In an attempt to lessen the fears of its monopoly power and gain wider support for the cellular proposal, AT&T announced that it would not manufacture cell phones, and that Bell Labs would help cell phone manufacturers create the new radio designs these phones would require. This concession gained the support of some manufacturers, but important concerns remained." A large new market for cell phones might develop, but "a standardized cell phone would attract new international competition" (Frenkiel 2010, 16). The government's breakup blew up AT&T's plan to develop the smartphone. It would likely have been AT&T, not Apple, that was the smartphone's market leader, had there been no regulation.

In the absence of AT&T, other mobile manufacturers became active. The FCC approved commercial cell phones on a regular basis in 1982. With the breakup of the Bell system and various FCC rulings that allowed consumers to own their own phones, the consumer telephone market was expected to grow from a $400 million business in 1982 to close to $1 billion by 1995 (Schrage 1983). By 1984 there were 25,000 users of the new phones (AT&T Archives 2011). Once Motorola introduced the world's first commercial handheld cellular phone, the DynaTAC 800X, the number of cellular subscribers surpassed 5.28 million in 1990, and the world's first commercial text message was sent in 1992 (US Census Bureau 2012; *Boston Globe* 2014). Monthly subscription fees also continuously declined, down from $86 to less than $40 in 1998 (fig. 2).

After AT&T failed to develop the smartphone in the early 1980s, the smartphone did not show any significant progress until the early 1990s, when a number of cellular phones emerged, as companies and consumers became fascinated with the touch-screen system and how to infiltrate it into common use. Even if smartphones are considered state-of-the art technologies in 2015, the first ever device of this kind dates back to 1994, when the concept of the "smart" phone in the home telephone sector that was advanced in the 1980s became implemented in the cellular phone industry. Of course, later operating systems should be credited with establishing the foundation for the smartphone revolution of the 2000s.

As several early smartphones, including the failed AT&T smartphone, prove, it is not unusual for new ideas and technologies run into difficulties during their diffusion due to several barriers, including lack of required concepts, funds, and use. As a consequence, some new ideas and technologies are eventually taken up by potential users, while many innovative products disappear before they can turn into valuable services and goods. A new type of smartphone, the IBM Simon, was unlike its predecessors, however, in terms of its actualization of the plan. Although it was not commercially successful, it was relatively well diffused over a couple of years. Its adoption at least for a brief period, as well as its advanced functionality, makes the IBM Simon the first smartphone.

The Emergence of the First Smartphone: IBM Simon

Having the size and shape of cell phones (which were much larger than contemporary smartphones), the IBM Simon, distributed by BellSouth, became the first-ever smartphone device. The IBM Simon could be prop-

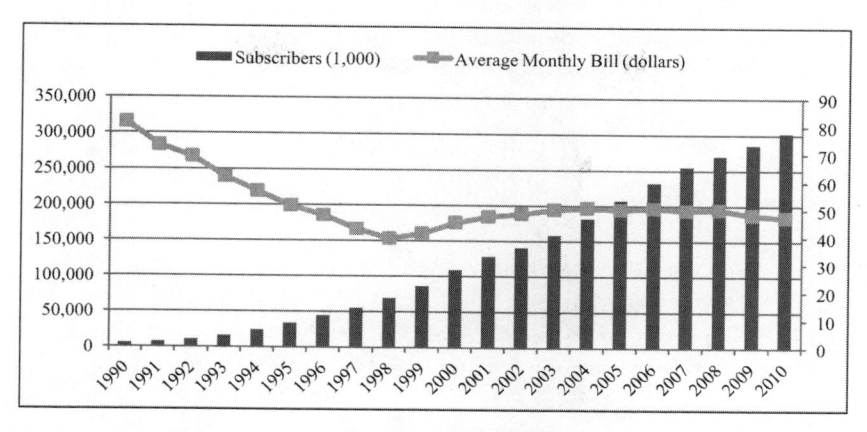

Fig. 2. U.S. Mobile Telecommunications Indicator, 1990–2010. (*Source:* Data from U.S. Census Bureau 2012, 720.)

erly referred to as a smartphone, and the launch of Simon gave birth to the modern smartphone era (Sager 2012). Several significant elements drove the growth of the first smartphone, and it is necessary to understand the major elements in the invention and growth of this first smartphone (Hughes 1994; Volti 2008). In other words, it is critical to historicize the social, political, and economic forces that exerted their influence when the IBM Simon emerged.[5]

American inventors and companies created the first-ever smartphone, as with many other technologies and technology cultures. "The IBM Simon replaced the usual telephone keyboard with a sensitive screen, and integrated PIM (personal information manager) applications and data communication features" (BellSouth-IBM Simon 1994). Simon was Bell-South Cellular's intriguing new so-called personal communicator. Developed largely by IBM, Simon was essentially a cellular phone with the electronic innards and LCD face of a PDA. As such, Simon not only served as a portable phone, it also filled the cellular airwaves with data. It could send and receive e-mail, faxes, and pages. And while each of these avenues had its limits, Simon was easily the most communicative PDA to appear thus far (O'Malley 1994).[6]

Simon was not the only communication device to pursue the integration of functions. However, the Simon showcased a totally different idea of what a communicating device could be: instead of integrating a modem in a PDA, it integrated a PDA and a modem in a cell phone (BellSouth-IBM Simon 1994). When it was on the market, observers still called it a PDA, not a cell phone. For example, in the section "Product of the Month" in *Telecommunications* (1994), an article was titled "BellSouth Cellular / IBM Release Simon PDA"; it commented that "the PDA can be used to return calls from pages by the push of button. . . . The Simon PDA measures 8 × 2.5 × 1.5 inches." Thus, the first cell phone to incorporate PDA characteristics was a model developed by IBM in 1992, before anyone called a cell phone a "smartphone." Rich Guidotti, who was BellSouth Cellular's product manager, said, "Simon represents the first real personal communicator because it was designed to be a cellular phone—a communication device—first, and a computer second" (Business Services Industry 1993). The user's manual (1994, 1) also emphasized the role of communications. In its introduction of Simon, the manual says:

> Welcome! You now have total personal communications—including your cellular phone—in one small, hand-held, mobile device. Simon has everything you need: cellular phone, fax, E-mail, pager, paperless notepad, address book, calendar, and calculator. And it's wireless! So it works where you work, goes where you go.

Guidotti recently said in an interview with the *Orlando Sentinel* (2013) that "the reaction among those attending that 1993 Orlando show was overwhelming: it was very similar to when Apple introduced the iPhone. People were in awe because it was a revolutionary device. We saw great potential for the business person or a real estate agent who is on the go." Paul Mugge, former IBM engineer, states:

> Before there was the Palm, the Blackberry, the Samsung Galaxy S3 or the iPhone 4S there was the Simon Personal Communicator. On November 2, 1992, IBM debuted the first "smartphone" prototype under the code name "Angler." The prototype was the combination of a cell phone and PDA. It gave the user the ability to make and receive phone calls and emails and receive facsimiles and pages. (Center for Innovation Management Studies 2013)

IBM engineers first showed a working prototype at the 1992 Comdex computer show in Las Vegas. Frank J. Canova Jr. is the IBM engineer who came up with the original concept for Simon. In the early 1990s, he was thinking that chip-and-wireless technology was becoming small enough to put in the palm of your hand. According to James Lewis, who worked on a development team, Frank also had an idea to make a touch screen

(Bradner 2011). He described his concept to colleagues, including his boss, Jerry Merckel, who was on an industry task force working up specifications for a new device (the PCMCIA card) that could plug into a laptop computer for extra memory. Merckel realized the cards could be used to launch other apps or services for Canova's phone. Mugge, who was director of the Florida Research Lab, put together a small team of engineers including Canova and Merckel to explore ways to use ever-smaller, more powerful electronics to build new products (Center for Innovation Management Studies 2013).

With the help of Frog Design, "IBM decided 14 weeks before Comdex that it wanted to display the device at the trade show" (Sager 2012). Paul Lasewicz (2012), IBM's corporate archivist, notes that in 1992, "a small team of IBM engineers from the IBM PC company's advanced technology group in Boca Raton was assembled to create a new mobile device that combined a computer with a cell phone. It was called a personal communicator—we know it today as a smart phone." Canova and other engineers worked 80-hour weeks, including weekends, right up to the last day. However, IBM was not sure it wanted to be in the phone business. Some internal critics called it a "World War II walkie talkie." Nevertheless, the effort moved forward, and the prototype at Comdex was displayed. Canova recalled, "Here I was, talking to someone with access to my calendar, email, and much more, with only a phone in my hand. For the first time, no computer was needed. That simple moment is when I realized the world was about to change" (Sager 2012).

Unlike the technological side, IBM had no confidence about the marketing of Simon. After seeing the prototype, BellSouth Cellular wanted to take action, and it was BellSouth that came up with the name Simon. "BellSouth Cellular teamed with IBM to turn it into a commercial product. . . . The two companies hold 11 Simon-related patents" (Sager 2012). One of the marketing managers had seen his kids play with the popular electronic memory game Simon, which asked players to repeat a series of tones that got progressively more difficult in order to win. The name Simon also suggests the game Simon Says, as in "Simon says simplicity." The Simon Personal Communicator came out in 1993. However, Simon was not ready for its scheduled release in May 1994, because IBM was wrestling with the device's short battery life (Sager 2012).

The Simon did not last long on the market. The device was ahead of its time, but by today's standards was huge, heavy, and expensive, both in purchase price and in costs for service subscription (BellSouth-IBM Simon 1994). The weight of Simon was 18 ounces, and it retailed for $899, comparable to $1,451 in 2016 dollars. Heavy data users had about 60 minutes before they needed to recharge the battery—as little as 30 minutes in areas with poor cell coverage. Simon sold approximately 50,000 units, and by early 1995, it was off the market. "IBM decided not to pursue the business. BellSouth also put money into improving its own communications network" (Sager 2012):

> The concept got the go-ahead but as advanced as the device was, it did have its setbacks. There was not sufficient Internet connection or bandwidth to support such an advanced device. Simon also had severely short battery life issues. Even after reworks in the software, the most feasible solution was to provide a second battery for the Simon Personal Communicator. Even after a lackluster performance, IBM had plans to produce a follow-up device to the Simon Personal Communicator called the Neon. However, the Neon never made it to market. The Simon Personal Communicator was ahead of its time. In fact the Simon was so far ahead of its time that there was not enough of a technological ecosystem in place to support it. Pioneers and innovators sometimes face roadblocks but in reaching those roadblocks they create paths for the great minds that will come behind them. (Center for Innovation Management Studies 2013)

Because of the heavy investment required for innovation, the company's lack of marketing experience, and insufficient confidence among managers, IBM was very cautious about taking on all the potential risks.

As is well documented, "There is a well-entrenched marketing rhetoric that says innovation is the lifeblood of businesses: that if they don't invest in R&D, push into new markets and maintain a new product development strategy, they will not survive. By being ahead of the competition with new products, companies gain first-mover advantage: they capture market share, establish a premium, and are seen as pioneers and able to build brand loyalty" (Bainbridge 2013). Yet if we think of some of the biggest, most successful global brands, such as Google, Amazon, and Facebook, the conventional wisdom on first movers looks not always sage (Bainbridge 2013). Followers can learn from pioneers' mistakes, and this late-mover advantage allows a company to see whether there is a market worth entering and judge consumers' tastes (*Marketing* 2013). IBM, although it was the first mover, chose the path of the late mover.

BellSouth did not plan a national ad campaign, instead "relying on its local cellular operators and retailers" to showcase Simon. "As a smart cellular phone, Simon should have had a better opportunity for success than other early personal communicators" (Johnson and Fitzgerald 1994). BellSouth, however, was reluctant to spend money because of the uncertainty about the market and the technology. As Sahin (2006, 14) points out, "Uncertainty is an important obstacle to the adoption of innovations." Paul Saffo, a technology futurist and director of the Institute for the Future in Menlo Park, California, said at the time, "This is a much bigger revolution than the personal computer. It's just going to take time" (Johnson and Fitzgerald 1994). Unluckily for both IBM and BellSouth, the Simon did not work in the available market. "IBM was hemorrhaging money and people, losing $16 billion and over 100,000 jobs in the years from 1991 to 1993. In the end, technical limitations, product delays, a world-class corporate meltdown, and bad business decisions conspired against Simon," which resulted in its closing down (Sager 2012).

At a time when a laptop seemed pretty foreign to both consumers and businesspeople, a smartphone seemed just too out there. There was not a market or an ecosystem for devices like the Simon or Apple's Newton. Simon was expensive, and IBM could sell it only to enterprises: it was not a consumer device (Connelly 2012). In fact, a major difference between today's smartphone and the early smartphone is that the latter was predominantly meant for corporate users and used as an enterprise device; these phones were simply too expensive for general consumers. During this phase, smartphones were targeting corporations, and their features and functions were aimed at corporate requirements (Sarwar and Soomro 2013). "Although several new technologies, including Alexander Graham Bell's original phone, found a large enterprise customer base, followed by much later general users, Simon did not follow a similar logic" (US Department of Commerce 1975, 783–84).[7]

Regardless of the failure of IBM's Simon in the market, the story of Simon tells "the timeless lesson of tech innovation: groundbreaking products require a rich ecosystem before the big idea can become truly useful or widespread. In this case, what was needed were fast networks, web browsers, and a whole lot of apps waiting to be pulled off the Internet. In the early 1990s, none of these were available. Phone networks were designed mostly for voice, not sending data. When Simon was conceived, a web browser had yet to be released" (Sager 2012). The first type of smartphone was not smart enough, although its influences on subsequent mobile technologies were huge. Timing and functionality, before the explosion of the Internet or fiber optics with tremendous bandwidth, caused the short life span of Simon (IBM Corporate Archives n.d.). The convergence of communication and computing for mobile consumer devices was required on "the evolutionary course to bring interoperability and to leverage the services and functions from each and every industry. In this process of convergence, smartphones were the leading devices taking the front end and playing the role of universal mobile terminal" (Zheng and Ni 2006; Nath and Mukherjee 2015, 294).

The Convergence of Smart and Phone in the Smartphone

The combination of the two words *smart* and *phone* in one word came with Ericsson's GS88 Panelope in 1997 (Stockholm Smartphone 2014). The smartphone product category was created by Stockholm Smartphone, although only 200 units were produced, as concept phones.[8] The lessons learned were used in developing the R380. Ericsson's R380 was the first phone to ship with an operating system (OS), Symbian, which (due in no small part to its open-source nature) would gain popularity around the world (McCarty 2011). The Stockholm Smartphone team placed a number of innovative smartphones on the market. The team invented another smartphone after IBM Simon, developed the touch screen, and was in business until 2010 (Stockholm Smartphone 2014). Ericsson presented the Smartphone R380e, the improved version of its widely acclaimed R380. R380e had new functionality, increased performance, and extended talk and standby times, as well as a new color, Denim Blue. Ericsson was out with a smartphone first when the company released the award-winning R380s in 2000 (Conabree 2001). As a combined mobile phone and personal organizer, the R380e was the ideal business tool leading to the mass production of the smartphone in the early 21st century.

Transnationalization of Smartphone Production

As Apple is a transnational telecommunications giant, the spread of U.S. technological culture in the realm of the smartphone has been much stronger than with other cultural products. As discussed, American ICT corporations developed the first smartphone; however, Japanese ICT firms were involved in the manufacturing process, which resulted in the growth of Japanese mobile industries in the early 21st century. Again, Simon was designed by IBM and marketed by BellSouth, but it was built by the Japanese corporation Mitsubishi Consumer Electronics of America (Johnson and Fitzgerald 1994). "Motorola, a supplier of cellular smarts for the prototype, passed when it came time to build the product, concerned that it would be helping IBM become a competitor. Mitsubishi therefore replaced Motorola and built the commercial product" (Sager 2012). Mitsubishi was developing a wireless personal digital assistant and handheld personal communicator to align with its projected cable TV interactive video network, which would extend remote service beyond the home and office; therefore, it was a logical choice to integrate Simon with its own cellular radio (Robertson 1994). Although it did not commercially succeed, IBM as an American-based transnational corporation, relied on the international division of labor in order to produce Simon.

Mitsubishi was not the only Japanese electronics company manufacturing the first smartphones. Almost at the same time, when AT&T and Eo Inc. developed a mobile phone similar to the IBM Simon, which was clunky and overpriced at $2,500, it was built by Matsushita Electric Industrial (Johnson and Fitzgerald 1994). "It was too big to slip easily into a briefcase or purse and too expensive to buy on a whim, and at first glance it resembles the head of a robot flattened by a steamroller. But the AT&T Eo Personal Communicator 440 is still an intriguing device, and it provided important wireless communications and computing tools for mobile executives" (Lewis 1993).[9]

The U.S.-Japan smartphone connection started when, in 1947, staff members at the Western Electronic division of AT&T made the first transistorized electronic components. Licenses were granted to a small Japanese company that wished to acquire the rights; the fee charged was a $25,000 advance against future royalties (Barnet and Cavanagh 1994, cited in du Gay 1997). While American ICT firms developed the prototype of the smartphone, Japanese ICT firms commercially made a few successful cell phones, comparable to the smartphone. Of course, Japan's partnership with global transnational corporations is not limited to the United States. As discussed, Sony Ericsson made GS88, the first mobile phone titled a smartphone. Sony Ericsson was a multinational mobile phone-manufacturing company, founded in 1987 as a joint venture between Sony and the Swedish telecommunications equipment company Ericsson. Sony acquired Ericsson's share in 2012 and the division became Sony Mobile Communications AB (Sony 2012).

Through its involvement in the early manufacturing processes of the smartphone in cooperation with these transnational corporations, Japan developed early commercial cell phones, comparable to the current smartphone. "In Japan in 1992, when new common carriers were accepted into the mobile communications industry, mobile communications consisted primarily of car phones. Although there were hopes for the future, at the time there were only about 1 million subscribers. Nobody anticipated the adoption rates that we see today. NTT DoCoMo was established as a spin-off of monopolist NTT." The subsidiary (DoCoMo standing for "Do Communications on the Mobile") was dedicated to wireless communications, including car phones and the nascent pager market (Kohiyama 2006, 62–63). In 1999, NTT DoCoMo released the first smartphones to achieve mass adoption within a country.[10] With 39 million cell phone subscribers and revenues of $39 billion in 2000, DoCoMo was certainly everywhere in Japan—yet the rest of the world knows it, and i-mode, only by reputation. DoCoMo—two-thirds owned by Nippon Telegraph & Telephone—invested in mobile carriers around the world, including AT&T Wireless in the United States (Rose 2001).

> The Japanese word for cell phone is *keitai*, which means "portable," and it's not hard to see why they were a bigger hit than home computers. . . . The first *keitai*, the so-called candy-bar models, had small black-and-white screens and were about half the size of Western cell phones. Now, with the advent of color and animation, the featherweight candy bars are giving way to slightly heavier, folding handsets with larger, high-resolution screens. They combine in one sleek device the functions of three separate gizmos in America: cell phone, handheld computer, and wireless e-mail receiver. . . . Technology is an expression of the culture that produces it, and Japan leads the world in consumer electronics because it's a society that places enormous value on convenience. (Rose 2001, 3)

What Japanese mobile industries experienced was too advanced to achieve a large market, as IBM had found out with Simon. Japan was also too satisfied with its own advanced technologies, and could not develop them furthermore.

The United States and Japan were not the only two countries to develop smartphones before the iPhone. Outside of Japan, Nokia 9000 Communicator really brought on the smartphone era. It was the first cell phone that could also be called a minicomputer (though it had limited web access). When opened, the longways clamshell design revealed an LCD screen and full QWERTY keyboard—the first on a mobile phone.[11] The Nokia 9000 Communicator was a cell phone that was a smartphone before the word was invented. It rolled some of the features of a computer into a phone, putting e-mail, web browsing, fax, word processing, and spreadsheets into a single pocketable device. Launched in 1996, the Nokia 9000 Communicator showed a company at the peak of its design powers (Baguley 2013).[12]

Of course, from our current perspective, the Nokia 9000 had several design limitations. First, the keyboard was very small, making it difficult for users to type with all ten fingers. Second, users had to navigate the screen by means of buttons, which are inferior to a pointing device. Third, the screen was small, limiting the amount of information it could display at one time. Last, it was rather bulky, making it uncomfortable to hold when used as a cellular telephone (Narayanaswamy et al. 1998, 60).

The next smartphone to achieve mass adoption was the Blackberry, which, because of its addictive nature, led to coining of the term *Crack-berry*. Research in Motion first released its GSM BlackBerry 6210 in 2003, and later released the Blackberry 7730, which featured a color screen (Ha-levy 2009). These new waves of phones allowed users to e-mail, fax, and make traditional calls, making them a must-have tool for the executive on the go.

As we have seen, manufacturers in several countries other than the United States have developed their own pre-iPhone smartphones, either individually or transnationally. Because these countries developed their own mobile phones and, later, smartphones, mobile technology in the U.S. lagged behind Europe and Asia until very recent years, although the U.S. entered the smartphone era ahead of other countries. New York City was a wireless underachiever in 2000, while Finland (and other countries) advanced mobile and smart technologies (Murphy 2000). Nevertheless, we have to admit the huge impact initiated of these first forms of the smartphone, which provided a foothold for later devices that contemporary society heavily relies on.

Conclusion

This chapter documented the early history of the smartphone, which is one of the most significant digital platforms in the 21st century that may be analyzed through social constructivism theory. Although our contemporary smartphone is not the bulky, limited, expensive device that emerged in the last century, and we may find it hard to imagine the original forms of the smartphone, it is crucial to understand the origin of the smartphone that so deeply influences our contemporary daily lives. In the information society, we now use a smartphone in which converge all kinds of features with mobility. Somewhat more than two decades have passed since the first smartphone—IBM's Simon—in 1994. The iPhone's predecessors are significant because, without them, the iPhone and Galaxy might not exist, and we would still be using feature phones in the 21st century.

The histories of the smartphone reveal several significant lessons. It is critical to analyze the history of the smartphone in the context of overall society, because social structures and processes affect both the invention of technologies and the choices of technologies, as social constructivists (Winner 1977; Mackenzie and Wajcman 1999; Volti 2008) point out. This was certainly the case in the pre-iPhone era. First, the growth of the smartphone was influenced by sociocultural elements such as prevailing communications policy and market conditions. While changing government policies in the early 1980s, resulting in the breakup of AT&T, deterred the emergence of the smartphone, corporate decisions in the early 1990s influenced the rise of the smartphone in our modern history.

Second, the uptake by users was not significant in the first stage of the smartphone because of its high price and monthly bills. As McKelvie and Picard (2008) point out, although the media space is today increasingly controlled by consumers, it was in the early 1990s controlled by ICT firms. It was a supply market, because companies that developed the first smartphones pulled out of the market, lacking experience and confidence. They were major customers of new technologies, which made them critical, not only in the invention of technologies, but also in the diffusion of these technologies.

Third, inadequate functionality played a key role in the delay of smartphone culture. The gadget was too big, people could not put it in their pocket, and batteries were not efficient. Since it lacked several functions characterizing current smartphones, including a digital camera, people did not buy this less desirable cell phone. The role of consumers was not comparable to the smartphone users of today, who eagerly wait for and buy new models.

Finally, but not least, the smartphone industry was transnationalized. A division of international labor, not between Western and non-Western countries, which is the current form of transnationalization, but between different Western countries, namely the United States and Japan, was a significant factor in the production of the early product, resulting in the growth of the smartphone in Japan. Meanwhile, some transnational telecommunications corporations that did not seize first-mover advantages, including IBM, Motorola, and AT&T, lost their greatest opportunities to be global leaders in the smartphone industry in the 21st century. As Bijker et al. (2012) point out, again, technological artifacts result from a complex interaction between technical capabilities and the interests and values of many individuals and groups. However, in the case of the smartphone, one also needs to add the significance of the interplay between a few countries, in particular the U.S. and Japan as major players. The relatively short history of the smartphone shows that some companies did succeed for a time and then declined and were forced to change their business models (Goff 2013). The rollout of the smartphone over the last 20 years has been impressive, and new variants will continue to emerge, as in the case of social media. The early smartphones, including the IBM Simon, definitely left a well-lit avenue for the contemporary smartphones that many of us cannot seem to live without.

PART II

Political Economy of the Smartphone Systems

3

Neoliberal Transformation of the Mobile Telecommunications Systems

Korea has rapidly developed mobile communication as a significant area of information and communications technology. As Korea has advanced its own ICT since the early 1990s, the country has invested in mobile telecommunications because of their significant role as new growth engines for the economy. Smartphone technologies have become among the most important in the Korean ICT sector since late in the first decade of the century, and perhaps surprisingly, Korea has become a global leader. The belated launch of the iPhone in Korea in November 2009 and its quick penetration nationwide have caused a ripple effect (or iPhone effect), created "a chain reaction in the mobile carrier-centered market" (Y. Lee 2010; D. Lee 2012), and the country has shifted its focus from mobile phones toward touch-sensitive smartphones.

The smartphone-led revolution is very complex, of course, because several socioeconomic and cultural factors have been involved. In particular, the Korean government has executed two different approaches during the transformation, consequently creating "smartland Korea." On the one hand, the Korean government has expanded its neoliberal telecommunications policies, including market liberalization and deregulation. Korea has advanced its own mobile technologies and industries since the mid-1990s with corporate-friendly telecommunications policies in the name of neoliberal reform. On the other hand, the Korean government has strongly initiated the growth of the mobile telecommunications sector, as the government has encouraged the growth of the Korean national economy through top-down and export-led economic policies since early 1970s. Therefore, it is crucial to understand the nature of the government involvement in the telecommunications industries, both wired and wireless, in order to comprehend the broader political and economic situations that Korea has faced.

This chapter historicizes the recent growth of smartphones and relevant apps as a transition toward another juncture of the Korean ICT scene, because the remarkable development of the smartphone needs to be analyzed within the overall context of the growth of mobile technologies. I first discuss the ways in which Korea has developed various ICTs. I study Korea's smartphone technologies and policy issues as part of the continuous development of the local telecommunications system. The main role of the Korean government in reshaping the mobile telecommunications system is investigated by examining how the government formulated its wireless telecommunications policy, combining neoliberal reform and Korea's developmentalism. The chapter then maps out how Korean mobile telecommunications industries have been altered by domestic political-economic factors as well as the current transformation of the global telecommunications industry. I also investigate how the Korean government and telecommunications industries protected the domestic mobile industry right before the introduction of Apple's iPhone, which has resulted in the growth of the domestic smartphone sector.

Transformation of Korea's Economic Model: From State-Led Developmentalism to Neoliberalism

Korea's political economic milieu has swiftly changed along several key dimensions since the early 1960s, when Park Chung-hee (Pak Chŏng-hŭi) took over political power via a military coup. Korea suffered economic hardship after the Korean War (1950–53); however, under the Park regime, significant political and economic changes were made. Most of all, the Park regime established the Korean economic model, which became a central part of the regime's legitimacy from the early 1960s onward. At that time, Korea initiated three major economic plans for growth: state-led industrialization, an export-led economy, and consequently nurturing chaebol (Korean conglomerates) (Amsted 1989; Hart-Landsberg 1993; J. H. Park 2002; Jin 2011; Wade 2004; Larson and Park 2014). There is no doubt that "the state, rather than market forces, especially played the key role in structuring the Korean economy and shaped growth. Top-down economic directives and regulations dominated, all under the rubric of a slogan calling for economic development" (Heo and Kim 2000, 494). As Joseph Stiglitz (2003,

94) argues, "East Asian governments, in particular, the Korean government, undertook major responsibility for the promotion of economic growth through a state-led and export-driven economic model." Korea's "political institutions are authoritarian rather than democratic, and state-corporatist rather than social-corporatist or pluralist" (Wade 2004, 306). This implies that the Korean government took a major role in the era of development in the 1970s and 1980s. A strong tradition of bureaucratization, centralization, and, more specifically, state-led developmentalism is an explicit feature of public policy in Korea (Oh and Larson 2011).

While there are several areas in which one can trace the outcomes of state-led developmentalism, the structural transformation of the industrial sector has been exemplary. The government promoted Korea's industrial transformation, from agricultural to light industry, such as the manufacture of shoes and clothing, in the 1970s to heavy industry, including shipbuilding and electronics, in the 1980s. Then, in the mid-1980s, Korea directed its industrial structure toward a knowledge-based economy centered on the development of ICTs. "The driving force of the government behind industrial change was apparent in the 1970s, and for that matter, in the mid-1980s. For example, the government had been promoting the domestic electronics industry as one of the nation's key sectors for almost two decades. . . . In the field of semiconductors, the government was taking the lead in the 1980s" (Amsden 1989, 81–83).

Against this backdrop, the Korean government has on a large scale supported information and communication technologies, having appropriated the ICT sector for sustainable economic growth. Korea has developed ICT systems since the early 1980s, and the rapid growth of ICTs accelerated in the mid-1990s when the government initiated its drive to develop ICTs for sustainable economic growth. For example, as a major part of these ICT policies, the government transformed the Ministry of Postal Service into the Ministry of Information and Communications (MIC) in 1994.

One of the most significant policy changes in the ICT sector came from the civilian government. In 1995, the Kim Young Sam (Kim Yŏngsam) government (1993–98) launched an information highway project designed to transform Korea's industrial system, which depended on heavy and chemical industries, toward a more ICT-driven industrial system. The Kim government authorized "a $45 billion project that would provide a variety of advanced telecommunications and multimedia services, including wireless PC communications, video on demand (VOD) and home shopping by 2015" (H. M. Chae 1997, 8).

For this purpose the government set up the comprehensive plan for Korea Information Infrastructure (KII) in March 1995, the goal of which was to construct an advanced nationwide information infrastructure consisting of communications networks, Internet services, application software, computers, and information products and services. The KII project was aimed at building high-speed and high-capacity networks through market competition and private sector investment, as well as government policies (Lee et al. 2003, 84; K. S. Lee 2011; Jin 2011). The KII was also a result of direct and indirect global pressures; it was a strategy to enable a nation-state to survive in the digital era of global capitalism. At the beginning of the 1990s, "The Korean government was forced to decide whether to remain a member of the second-tier countries under the digital mode of capitalism or to find a way to make a leap forward" (K. S. Lee 2011, 54). The government chose transforming the nation "from a labor-intensive economy" to "a knowledge-based economy" as the foremost goal of state affairs, as Kim Young Sam addressed in his New Year's speech in 1995 (D. Lee 2012). The government's role in this massive infrastructure project included both implementation of the public segment of the project that networked schools and public organizations, and the promotion of informatization to ensure demand for broadband (Oh and Larson 2011; Larson and Park 2014).

The ICT-driven economic policy has changed the map of Korea's national economy. For example, as indicated in figure 3, annual exports in 1997, right after the introduction of KII, were recorded at US$136.1 billion, and the ICT sector accounted for 22.9% (US$31.2 billion). In 2014 annual exports increased to US$573.1 billion, and the ICT sector consisted of 30.3% (US$173.9 billion) (Ministry of Science, ICT and Future Planning 2014b; 2015b). This means that the role of the ICT sector has significantly grown over the last 17 years. What is interesting, though, is that the Korean economy has changed since its adoption of neoliberal policies. Ironically, the Korean economic model has nurtured the rapid adaptation of globalization, because Korea, with its export-oriented approach to development, was highly exposed to the global processes set in motion by the arrival of digital networks and the Internet era (Larson and Park 2014).

Neoliberal transformation in Korea started in the early 1980s. At that time, the newly established Chun Doo-Hwan (Chŏn Tu-hwan) regime vehemently promoted neoliberal reform in several areas, including the economic system (e.g., banking system) and the telecommunications industry. The new military regime arguably "abandoned the old idea of a developmental state," which the Korean government had pursued, and

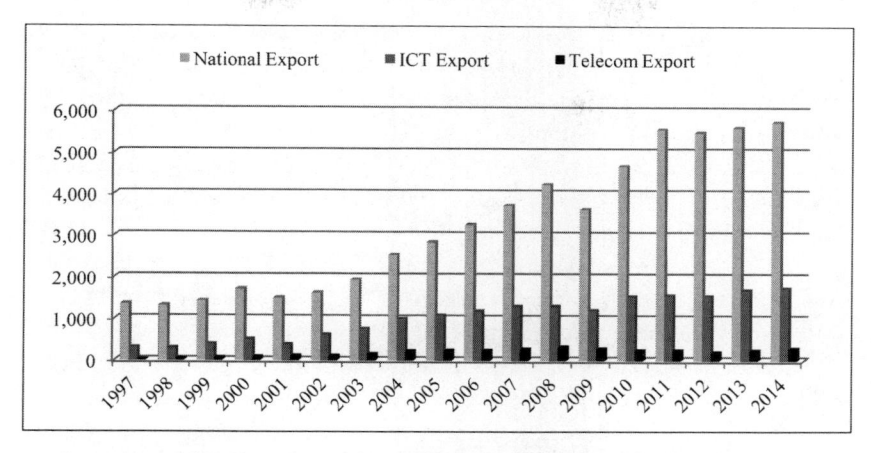

Fig. 3. Annual ICT Export Data (unit: millions). (*Source:* Data from Ministry of Science, ICT and Future Planning [Mirae Ch'angjo Kwahak Pu] 2014b, 2015b.)

"decided to ally with neoliberal economists, who had been influenced by the emerging ideologies of marketization and privatization" (Y. T. Kim 1999, 445). The Korean government has pursued neoliberal reform in the telecommunications sector, as have many other countries. Beginning in the early 1980s, the US government, alongside several transnational corporations, forced the majority of governments to undertake neoliberal reforms, including the liberalization and privatization of the telecommunications market and system, opening it to Western corporations, of course, mainly for American firms, intensifying the dominance of Western countries. Many developing countries, including Mexico and Chile, have, arguably, adopted US-led neoliberal globalization in order to survive in the globalized network society (McChesney and Schiller 2003). The Kim Young Sam government also intensely pursued neoliberal economic policy, meaning the actualization of a small-government scheme in order to guarantee maximum profits for the private sector (Friedman 2002). Right after the 1997 economic crisis, the worst in Korean history, the Kim Dae Jung (Kim Tae-jung) government had to liberalize the country's market, and the 1997 economic crisis became a turning point in foreign direct investment (FDI) in Korea.[1]

Under these circumstances, the Korean mobile telecommunications industries, including both service providers and handset makers, have gone through a series of neoliberal reforms, including deregulation, liberalization, and privatization. The Korean government began restructuring the mobile telecommunications industry because of increasing demand for participation in the market from both domestic actors, including a few chaebol, such as Samsung, Hyundai, and SK, and international forces, including international organizations, transnational corporations, and the US government (Jin 2011). As had many other countries, Korea initially developed its telecommunications system in the public sector, while in the U.S. the telecommunications industry was a private, corporation-driven sector. Nonetheless, states other than the U.S. have always been "crucial and necessary players in the formation of successive telecommunications systems" (McChesney 2008, 307).

Unlike the wired telecommunications industry, which operated in the public sector, the mobile telecommunications industries have been a private sector-driven system since inception. In Korea, mobile phone service was introduced in 1984,[2] and it was monopolized by SK Telecom before Shinsegi Telecom launched its services in 1996 (K. S. Lee 2011; Bae 2014). Since the early 1990s, the Korean government has pursued neoliberal telecommunications policy in the wired telecommunications sector; it had already pursued a neoliberal telecommunications policy in the wireless telecommunications sector in the 1980s.

However, the government delayed the liberalization process of both the broadcasting and telecommunication markets because of their significant influences in Korean culture and the national economy. While the Korean government had no choice but to liberalize the market, it maintains, at least partially, if not entirely, its state-led developmentalism in these crucial areas for the national economy and culture.

State-Led Developmentalism in the Mobile Telecommunications Systems

The Korean government has continued and even intensified its primary role as the major player in the telecommunications sector in the midst of neoliberal globalization, because the government understands the importance of telecommunications as the primary player in the ICT-driven economy. The government has supported the telecommunications sector with its legal and financial arms, while enacting neoliberal reform, including liberalization and deregulation (C. Yoon 1999). Under this complicated but still favorable environment, the mobile telecommunications market surged as one of the biggest sectors in the overall ICT market. In tandem with the growth of early mobile technologies and industries, the

Ministry of Information and Communications advocated the mobile system based on the Code Division Multiple Access (CDMA) digital technology, which has become a standard form for the mobile telecommunications industry. It pursued the indigenous CDMA development project beginning in the late 1980s (MIC 1994, cited in Jho 2007).[3]

The successful launch of a digital CDMA mobile phone service in 1996 spurred cellular growth (National Computerization Agency 1998). In the early 1990s, the global wireless telecom market evolved from an analogue to a digital system with the availability of digital mobile technology. Korean telecommunications firms could either adapt to the digital technology by importing the products from Time Division Multiple Access (TDMA) companies for second-generation technology or by commercializing CDMA on their own for third-generation technology. "Commercial TDMA service was already in operation in the European mobile market, but European firms were not willing to share it with Korean manufacturers, and Korea decided to commercialize CDMA technology" (Jho 2007, 129).

Korea was the first country to commercialize CDMA technology, although it borrowed the basic technology from Qualcomm, a US company. At the time, "As the market for mobile Internet services grew, demand for higher capacities of data transfer in mobile communications increased. The third-generation of mobile technology was ready to cater for this demand and two alternative technologies, WCDMA (wideband code-division multiple access) and CDMA 2000, competed for market share" (Lee and Han 2002, 161–62).

Succeeding Korean governments—the Kim Dae Jung government (1998–2003) and the Roh Moo Hyun (No Mu-hyŏn) government (2003–2008)—continued the ICT-led economic policy, considering ICT to be of the most important forces that could drive an escape from the 1997 economic crisis, as well as a way to foster growth in the national economy, while emphasizing the importance of ICT for attracting foreign investment. In 2004 the Roh Moo Hyun government developed a plan to nurture the mobile phone industry, which was already one of the most advanced in the world. Korea aimed to secure not only growth but also leadership in the global mobile telecommunications sector, taking full advantage of the country's solid domestic infrastructure. Its mobile policy revolved around "u-Korea," or ubiquitous Korea, referring to "a futuristic telecommunications network that would be pervasive enough to offer uninterrupted access to the Internet" and fixed-line and mobile networks at any time, from anywhere. Pushing for a grand vision, the government designated the sector one of the 10 future-oriented industries (S. J. Yang 2003).

As part of this plan, the Roh government launched the "IT839" strategy, a strategic vision for the ICT industry with the aim of achieving GDP of $20,000 per capita. It is expected to shape the future of the ICT industry and is laying the foundation for new growth that will lead to ubiquitous ICT (Chin and Rim 2006). The strong growth exhibited by the ICT industry prior to this plan has been attributed to the government's strong policy of informatization. In particular, "the rapid development of digital technology triggered the fast integration and convergence of voice/data, telecommunications/broadcasting, and wired/wireless" (Chin and Rim 2006, 32).

In February 2006, the government readjusted IT839 to reflect changes in technologies, markets, the environment, and policy priorities. This updated strategy, u-IT839, put special focuses on eight services, three infrastructures, and nine new growth engines (Sung 2009).[4] Among these, the growth of mobile communication was the most significant, and the government planned to establish wireless portable Internet networks and create demand for wireless multimedia services, based on a world-class broadband infrastructure (Ministry of Information and Communication 2004).

When the Korean government initiated the development of ICTs, including telecommunications industries, there were several significant players, both domestically and globally, who intensely pursued reform of the Korean mobile telecommunications market. These actors asked the government to undertake dramatic neoliberal reform in order to meet global standards, meaning they demanded massive deregulation and liberalization measures. Notable demand came from foreign forces, among them foreign value-added service (VAS) suppliers, international organizations, including the World Trade Organization (WTO) and the International Monetary Fund (IMF), and transnational telecommunications corporations. The launch of WTO negotiations on basic telecommunications services in May 1995 compelled the Korean government to expedite liberalization as well as privatization in telecommunications, as in other sectors (Jin 2006). The major driving force for the opening of the Korean telecommunications market was the US government, which had a disproportionately heavy influence on telecommunications reform, as will be discussed later in the case of the Wireless Internet Platform for Interoperability (WIPI), as Korea was a lucrative market and a large trading partner (Larson 1995, 14; Jin 2011).

Internally, chaebol were increasingly interested in entering the mobile telecommunications businesses as suppliers, main users, and service providers. Among the major conglomerates, SK first entered in the mobile service industry in 1994 through the privatization of Korea Mobile Telecommunication (KMT), a subsidiary of Korea Telecom (KT) for mobile business, and expanded its business to other telecom business domains. As is detailed in the following section, chaebol massively invested in the mobile telecommunications service sector, attracted by their equipment production companies. In the manufacturing area, a handful of Korea-based transnational corporations, including Samsung Electronics and LG Electronics, produced handsets and became top mobile gadget makers in the world. Korea's chaebol turned out to be primary actors in the global mobile telecommunications area. As Dan Schiller (1982, 89) pointed out in relation to the American framework, conglomerates as "business users desire to participate in and have an influence on crucial telecommunications decision-making as they are the main users and beneficiaries."

The Korean government has had no choice but to adopt neoliberal reforms in the wireless telecommunications sector as in many other fields; however, instead of giving up its role, from the beginning of the neoliberal era the Korean government has developed its own unique neoliberal regime—meaning the government has advanced its leading role in the telecommunications sector, although it has cooperated with global forces, including the US government and international agencies, and TNCs in the name of the growth of national IT industries and the national economy.

Neoliberal Transformation of the Mobile Phone Service Market

Mobile telecommunications service companies have grown substantially in value over the last two decades. The government has pursued a neoliberal approach to the market, which has resulted in the concentration of the market in the hands of a few major players. In Korea, again, mobile phone service was introduced in 1984, and it was monopolized by SK Telecom before Shinsegi Telecom launched its services in 1996. KT Freetel, a subsidiary company of KT; LG Telecom; and Hansol.com also joined the market in the late 1990s, and the mobile phone service market grew under the relatively well-balanced competition between the five companies. Although SK Telecom (henceforth SKT), the largest mobile phone service company, accounted for 41.5% of occupancy rates in the market in August

1999, the other four companies together comprised 58.5%: KT Freetel (KTF) (18.8%), Shinsegi Telecom (14.3%), LG Telecom (14%), and Hansol PCS (later Hansol.com) (11.4%) (*Korea Times* 1999).

In the early 2000s, the mobile service market changed with the corporate convergence of several mobile phone service companies. In 2000 KT acquired Hansol.com, which was suffering from financial deficits, for its subsidiary company KTF (S. J. Yang 2002). The leading mobile phone company, SKT, also acquired Shinsegi Telecom in 2002, when POSCO, the largest shareholder of Shinsegi Telecom, sold off its shares to SKT as part of its move to concentrate on its core steelmaking business (K. H. Yu 1998). Following the two major mergers and acquisitions, the market share of SKT in the mobile phone service market soared to 56.8% in June 2003, up from 41.5% in 1999, increasing its dominant position in the market (*Chosun Ilbo* 2003, 24).

Furthermore, the government has approved a series of mergers in the name of competitiveness in the era of media convergence. First, SKT acquired Hanaro Telecom, a broadband service provider, in February 2008 by purchasing 38.9% of the stake from two foreign investors, AIG and Newbridge Capital, and their partners, following deregulation that would allow telecom companies to offer bundled services (*Forbes* 2007). The goal of SKT was to absorb the second largest broadband Internet subscriber base (369 million, 25.2%) and a landline subscriber base (200 million, 8.6%), and acquire the enhanced capability to provide IPTV (Internet protocol television) service. Rival companies such as KT and LGT expressed their concerns that SKT's mobile market dominance would spread into broadband and fixed-line services. Despite these negative opinions, the Korea Communications Commission (KCC), established in February 2008, approved the acquisition, and included conditions such as nondiscrimination in the wholesale market, in bundling of services, and in the wireless Internet market (Ministry of Information and Communication 2008).

Soon after, in March 2009, a merger between the country's biggest telephone company (KT) and second largest wireless carrier (KTF) was approved by the KCC (T. Kim 2009a). Considering the resistance of the other players, "the KCC ruled that KT should make its telephone and broadband networks more accessible to rival telephony carriers and Internet companies" (T. Kim 2009b). In December 2009, the KCC also approved the merger of the LG group's three telecommunication operators—LG Telecom (mobile operator), LG Powercomm (Internet service provider: ISP), and LG Dacom (fixed lines), which resulted in the launch of the integrated entity LGU+ in July 2010.

Telecommunications firms in Korea have been seeking various ways to boost profits in the one of the world's most saturated and fiercely competitive markets, by developing converged products and bundling fixed-line, mobile, and Internet services. Under the approved merger plan, LG Telecom, Korea's smallest mobile operator by revenue, absorbed fixed-line operator LG Dacom and Internet service provider LG Powercomm (J. Lee 2009). Korea's mobile telecommunication market has become oligopolistic in structure through continuing mergers and acquisitions, now consisting of three telecom corporations. As of December 2015, the market share of SKT was recorded at 47.1%, followed by KT (26.2%), and LGU+ (18.9%). Others (7.7%) were operated through mobile virtual network operators (MVNOs) (Ministry of Science, ICT and Future Planning 2016).

There is no doubt that this market reconfiguration was the result of neoliberal telecommunications reform. As Cheol Gi Bae (2014, 106) wisely observes, in Korea wireless telecommunications industries had been "increasingly liberalized in accordance with neoliberal ideas since the 1980s. Neoliberal philosophies were the basis for the dismantling of "public monopolies" in the Korean telecommunications industry through the introduction of competition." The proponents of neoliberalism, of course, mainly the business sector, believed that lessening government intervention would create viable markets and increase prosperity by increasing competition, both nationally and globally. However, "ironically enough, the decreasing power of regulators accelerated the consolidation of the telecom market and helped chaebol to shape an oligopolistic market structure" in the mobile telecommunications sector (Bae 2014, 106). These mobile telecommunications service providers have closely worked with handset makers who are also chaebol companies, such as Samsung and LG, to develop IT Korea. These service providers and handset makers have actualized the rapid growth of the smartphone era since 2009.

Wireless Telecommunications Policies in the Pre-smartphone Era

Introduction of WIPI

Korean wireless telecommunications policies have massively deregulated and liberalized the telecommunications industry in order to support corporate interests in the neoliberal era; however, the state-led economy has also required the government to actualize its own goals. Korea's stance encourages a nexus between the Korean wireless telecommunications policy regime and ICT industries, as in the case of the Wireless Internet Platform for Interoperability (WIPI) policy, which is a primary driver for the growth of domestic mobile technologies because it prevents foreign mobile corporations from infiltrating the Korean market.

In the pre-smartphone era, until 2008, there were "three stumbling blocks for foreign handset makers," including Apple, entering the Korean market. First, the local market was small, and therefore not seen as lucrative. Second, it was dominated by the world's second largest (Samsung Electronics) and third largest (LG Electronics) mobile sellers. And third, there are WIPI rules (H. J. Jin 2008a). Among these, WIPI shows Korea's unique wireless telecommunications policy, which implied a close relationship between the government and domestic corporations in the pre-smartphone era. WIPI is a middleware platform that allows mobile phone users to access the Internet with their handsets and download multimedia data, and after the government set WIPI as a mandatory standard in the Korean wireless telecommunications market, foreign mobile manufacturers that did not have WIPI could not be sold and used in the domestic market. In the early 2000s, the MIC stated that it had initially mapped out a plan to introduce a wireless Internet platform in a bid to reduce national mobile corporations' dependence on Qualcomm's CDMA technology (S. J. Yang 2003). However, WIPI functioned "as a legal and technical barrier" for foreign handset makers to enter the Korean market, and a few powerful transnational mobile phone manufacturers, such as Nokia and Sony-Ericsson, could sell only a few modified products, which resulted in their failure in Korea (G. Kim 2011).

The establishment of WIPI was a complicated policy measure to create because of severe conflicts among the major players, including the Korean government, foreign forces (e.g., the US government and transnational telecommunications corporations), and domestic corporations—both telecommunications service providers and handset makers. As a reflection of the state-led development shown in the growth of the Korean economy in the 1970s and 1980s, the MIC led the process of setting standards for WIPI. As data services were considered the next source of revenue after voice communications, mobile handset makers also made efforts to provide quality mobile applications by developing better platforms on which applications could run faster to ensure customer satisfaction. Mobile carriers in Korea considered platforms to be one of their critical resources for competing with others in providing mobile data services.

In Korea there were three mobile carriers, and each used a different platform. SKT used its own virtual machine (VM), while KTF used Qualcomm's BREW, and LG Telecom used Java. The incompatibility of these platforms created difficulties. "The existence of several different mobile platforms has been a headache for content providers. They had to develop different versions of the same content" (Oh and Larson 2011, 107). A standard mobile platform could "eliminate duplicate investments of content providers and allow them to focus more on their core business, content development" (Lee and Oh 2008, 666). The government intended to increase efficiency by eliminating duplicate investments and to encourage participation of potential entrants in the mobile Internet market by fostering a favorable environment for a unified platform (Lee and Oh 2008).

Another consideration was the need to cope with increasing royalties paid to foreign platform providers such as Qualcomm (BREW), and, therefore, the new standard was to be developed with domestic technologies. "The government also appeared to have thought that a new platform could become a global standard once it was ratified as the national standard and widely used in Korea" (Lee and Oh 2008, 667).

Under this circumstance, in May 2001, the MIC founded the Korean Wireless Internet Standardization Forum (KWISF) to lead the development of WIPI to create the standard platform (Electronics and Telecommunications Research Institute 2001). The establishment of a Korean standard encountered diverse forms of opposition from both foreign and domestic players. Since the original plan was to have a single standard mobile platform embedded in all the mobile phones used in Korea, foreign actors strongly resisted the introduction of WIPI, because foreign mobile phones could not use the standard. For example, Qualcomm's BREW for KTF could not be used in Korea, and, therefore, Qualcomm resisted the plan. A series of protracted negotiations were held between domestic forces, involving the KWISF and MIC, and foreign actors, including Qualcomm, the US Telecommunications Industry Associations (TIA), and US government agencies, including the US Trade Representative (USTR). The dispute lasted for a year, and "The U.S. called on Korea not to adopt a locally developed mobile Internet platform as a standard during bilateral trade talks" (T. Kim 2004).

The USTR especially insisted that the policy was a trade barrier aimed at expelling Qualcomm's BREW from the Korean market, something that was prohibited by the Agreement on Technical Barriers to Trade of the World Trade Organization (WTO). In April 2004, the USTR strengthened its stance by tagging Korea as one of its "key countries of concern" (T. G. Kim 2004). After a two-year stalemate, Korea and the U.S. finally found some common ground in their long-running wireless Internet platform dispute. In April 2004, the U.S. agreed not to take issue with Korea's unified mobile platform as long as the country allowed competition. The U.S. also agreed that the Korean government had the right to set a mandatory national standard for domestic wireless Internet platforms (T. G. Kim 2004; S. Y. Kim 2012). As a result, in the Korean mobile market the mandatory use of WIPI began in April 2005, although the government's mandated usage of WIPI was abolished in December 2008.

Abolishment of WIPI

WIPI did not last long. Several mobile carriers and civil organizations pushed to discard the policy, whereas LG Telecom and WIPI developer firms insisted on retaining it (J. H. Hwang 2008). The Korean government and a handful of corporations—primarily handset makers and wireless carriers—were unwilling to consider abolishing WIPI until 2007. However, the majority of players eventually agreed to end it because it lacked sufficient incentives for businesses. The abolition of WIPI allowed domestic consumers to use iPhones and Blackberrys. The process of abolishing WIPI was complicated and even more nuanced than its establishment.

Just as the government took a primary role in the establishment of WIPI, the government initiated its abolishment in the name of technological innovations and business interests. The Lee Myung-Bak (Yi Myŏng-bak) government (2008–13) created the KCC by consolidating the old telecommunications regulatory body (MIC) and the old broadcasting regulatory agency (Korean Broadcasting Commission, KBC). Taken together, "These sweeping changes gave great discretion to the private sector, both mobile service providers and handset makers, for a period of months while the new administration was being formed and preparing to pursue important policies" (Oh and Larson 2011, 108). The KCC became the exclusive authority for dealing with WIPI-related issues, and this new agency decided to lift a key mobile regulation starting on April 1, 2009, clearing away a barrier to the entry of iPhone, Blackberry, and other popular foreign handsets into the local market. The KCC said that the decision was aimed at expanding customer choice and responding to the global trend of opening mobile platforms. "The move will expand customer rights to choose handsets. It will also fuel competition, which will drive down handset prices in the long term," Cho Young-hoon (Cho Yŏng-hun), a KCC official, said at a news conference. He expected low-priced foreign handsets to make inroads into the domestic market, which was dominated by expensive handsets manufactured by Samsung Electronics and LG Electronics (H. Jin 2008b).

Meanwhile, wireless carriers gradually started to withdraw from the closed coalition of WIPI. KTF negotiated with Apple to sell the highly popular iPhone as the new WCDMA platform in the Korean market in August 2007 because the firm was eager to catch up with its stronger rival, SK Telecom (J. Cho 2007). At that time, "KTF needed to break down the SKT/Samsung alliance and enlarge its own market share. Samsung, a dominant mobile handset manufacturer, had provided its new premium phone lineup exclusively to SKT. Samsung later offered its already outdated lineup to KTF" (Bae 2014, 136). On August 27, 2008, in a public discussion about the mandatory policy for the installment of WIPI on mobile phones held at the YMCA in Seoul, KTF executive director Lee Dong-Won (Yi Dong-wŏn) said, "In 2005 when the WIPI mandatory policy began, the WIPI was created to solve the compatibility issue as well as the platform royalty issue. However, it is time to globalize the wireless Internet industry as a whole" (J. H. Hwang 2008).

Although it is now obsolete, the WIPI mandate had a significant effect in that it protected the domestic mobile telecommunications market. In particular, it gave domestic-based transnational telecommunications corporations a grace period to prepare for the impending entry of foreign telecommunications corporations, including Nokia and Apple, and helped the solid growth of Korea's smartphone industries. In fact, the main reasons for the delay of iPhone involved both the government and to a much larger degree Korea's chaebol. The government was involved because the WIPI requirement, originally introduced with good intentions, had by 2007 become, de facto, a nontariff trade barrier.

On the industry side, the mobile service providers, led by SK Telecom, feared "a disastrous loss of voice revenue" if the domestic market were opened up to the iPhone and other smartphones (Oh and Larson 2011). Korea's handset manufacturers, led by Samsung and LG, were in a different position, having established themselves as leaders in the global market on the basis of their CDMA-based feature phones. They appeared to be perplexed by the arrival of foreign smartphones, with the emphasis on software and the creation of a new ecosystem powered by applications (or "apps"). In retrospect, the delayed arrival of smartphones in Korea's market had a distinctly negative effect, "retarding progress and development of a new, higher value-added section in the nation's service market" (Larson and Park 2014, 354).

Interestingly enough, the major foreign smartphone maker, Apple, made no push to get WIPI eliminated, regardless of the fact that its entry into the Korean market might potentially shake up a handset market controlled by domestic manufacturers. Korea's regulatory requirements could have discouraged the iPhone's debut in the market. Other than the WIPI issue, the iPhone also ran into problems before launching in Korea due to concerns that "location services such as Google Maps would breach a privacy law" (*Independent* 2009). "Korea's regulatory investigation into the iPhone had hinged on laws concerning the access it would give to maps and global positioning applications—a sensitive issue in the country, which is conscious of both security concerns related to North Korea and to its own authoritarian past" (Oliver and Song 2009). The Korean market itself was relatively small and was notorious for foreign IT players. "The big problem for Apple was simple. Koreans had been more attracted to phones made by local consumer-electronics powerhouses Samsung and LG, both of which roll out scores of sleek multimedia handsets featuring leading-edge technologies every year" (I. Moon 2008). The Big Two together controlled nearly 80% of the Korean handset market at the time (I. Moon 2008; Y. Lee 2010). On the whole planet, there is one place U.S. brands are not likely to generate much buzz: Korea. The country is one of the most advanced mobile Internet markets in the world, and electronics companies have worked hard to make sure tech-savvy Korean consumers don't fall for foreign brands. "This is a place where Nokia is virtually absent. Google has struggled in Korea too." And while consumers in other countries have embraced the iPod, "most Koreans are just not that into Steve Jobs and the work of his Apple designers" (I. Moon 2008). However, domestic mobile service providers, KTF in particular, wanted to have Apple phones available in order to boost their market shares in the Korean market.

The powerful Internet capability of the iPhone is another issue that made Korean firms reluctant to introduce it (J. Cho 2007). Users of the iPhone could freely surf the Web via Wi-Fi wireless networks that were available in many public and private places, such as homes, offices, and even fast food chains, like McDonald's. As Cho Young-chu (Cho Yŏng-ju), CEO of KTF stated, "Such versatility is a minus for telecom firms who force subscribers to use their own mobile Internet services for higher fees." For that reason, none of them had officially expressed interest in the introduction of iPhone before. The telecom firms, including KTF and SKT, however, had been "put under growing pressure to change. Complaints had grown among tech-savvy consumers who were bored with the limited handset choices in Korea; Nokia, the world's largest handset maker, was not selling any of its products in the country." The government was also willing to open up the telecom market to new firms in order to encourage healthy competition in the field (J. Cho 2007). Sales of the iPhone were finally approved in September 2009 by telecom regulators, who said the services it offered would not encroach on privacy (C. K. Park 2009).

Dominant handset manufacturers such as Samsung and LG had benefited most from the WIPI mandate, and bitterly opposed the introduction of the iPhone made possible by abolishing WIPI. The prior closed market structure allowed both manufacturers to become key players in the global wireless telecommunications market (Bae 2014, 138). As Cheol Gi Bae clearly observes, their "virtuous circle of development involved (a) acquiring monopolistic positions in the domestic wireless handset market, (b) accumulating capital and technological know-how using the domestic market as a test bed, and (c) making inroads in the global market through low price and high quality. Samsung and LG were concerned that the iPhone would disrupt the existing markets and value network in the Korean wireless industry" (2013, 138).

Samsung and LG hoped to delay the iPhone's introduction and thereby obtain a grace period to improve their smartphone-making capabilities. It was an open secret that the government delayed the removal of the WIPI mandate in order to give Samsung sufficient time to prepare its own smartphone lineup[5] (Bae 2014; T. Kim 2009a). During the same public discussion hosted by the YMCA in 2008, officials of LG Electronics and LG Telecom attending the discussion said, "The market needs time to prepare. We are preparing for the post-WIPI era. But . . . the competitive landscape after the opening of the market to the global platforms [is worrisome]" (J. H. Hwang 2008).

Following a lengthy debate, commissioners of the KCC finally decided to retire WIPI, lifting what had effectively been a trade barrier for foreign electronics makers like Apple and Nokia. Shin Yong-sub (Sin Yong-sup), the director of KCC's policy bureau, said, "Mobile-phone operators have been required to use the WIPI mobile platform on their handsets, but considering global industry trends toward the use of general-purpose mobile operating systems, we concluded that there was a need to allow carriers the freedom to decide whether to use WIPI or not." KCC also stated that "consumers will also be able to choose from a wider variety of products and benefit from increased price competition from handset makers" (T. Kim 2008).

The WIPI policy was a window into the nature of Korea's neoliberal mobile telecommunications policy, because both the introduction and the abolishment of WIPI were the result of the conflicting relations among three major forces; the government, domestic capital, and foreign forces. The Korean government exerted strategic leadership over the domestic mobile telecommunications industry for the national economy, although it also needed to reflect demands for market liberalization. Foreign-based TNCs and domestic business continuously demanded the removal of the policy network surrounding WIPI, and it certainly influenced the government policy. However, this does not mean that the Korean government gave up its protective mobile telecommunications policy. The government evaluated the changing political-economic situation surrounding the mobile telecommunications sector, and it decided to dismantle WIPI because the government believed the domestic telecommunications industries were able to compete with foreign mobile corporations both globally and nationally. In the midst of strong demands by several actors, including several civil organizations representing potential iPhone users and a few telecommunications service providers, the Korean government withdrew from its original stance in the name of technological innovations and business interests. It is significant to understand that the abolishment of WIPI itself was a governmental decision, which means that the government again took a major role in the smartphone's history.

Post-iPhone Era: Smartphone Revolution in Korea

The phenomenal growth of smartphones only started right after Korea introduced the iPhone in November 2009. In Korea, a little bit of history is needed to understand the smartphone revolution. While Korea allowed the sale of iPhones in 2009, the concept of the smartphone started to circulate in the early 1990s. Of course, the production of cellular phones started in 1988 when Samsung Electronics became the first domestic firm to produce cellular phones. However, because of low brand awareness and inferior technology, Samsung was only able to capture 10% of the domestic market. It was only after 1993 that Samsung began to surpass Motorola, the previous leader in the domestic market, by investing heavily in technological capabilities and brand-name value. In 1996, LG Electronics joined the mobile handset market (Korea Trade Investment Promotion Agency n.d.).

As discussed, the rollout of iPhones was delayed for various reasons in Korea, including the WIPI policy. Heated campaigns by consumer groups accusing Korea of protectionism also played a role. The customers stressed it was ironic that the world's most wired country was keeping out Apple's smartphone, while it was available in less developed nations (Oliver and Song 2009). For many, smartphones served purposes beyond traditional mobile functions emphasizing telephony. They were transformed into handheld computers with Internet access. Smartphones became among the most innovative and cutting-edge technologies in the 21st century. As the handsets got smarter, the nature of the industry changed. "It would be less about hardware and more about software, services, and content, including applications (apps) for smartphones. As late as October 2009, using 3G mobile phones in Korea was universal, but only a little more than 10% of all customers purchased a data plan to use web-based services because of exorbitantly high data rates" (Oh and Larson 2011, 106–7).

In Korea, the era of the smartphone arrived relatively late, the Korean launch of the iPhone occurring in November 2009. As discussed, "The iPhone's debut was belated primarily due to both protectionist government telecommunication policy in the name of privacy (which blocked the sale of iPhones in the domestic market until 2009) and concerns about the future of business raised by telecommunications corporations" (P. Kim 2011, 261–62). However, the iPhone's quick penetration nationwide after its debut heralded the start of Korea's smartphone era. The government and telecommunications corporations were shocked to witness the astonishing growth of iPhone worldwide and started to rebuild their strategic plans. The rapid penetration of the iPhone both nationally and globally caused them to face the question of what they had missed technologically in the past and what they should do for the future (Korea Telecom 2010).

The iPhone effect was not minimal. Following the announcement of the pricing of the iPhone, about 40,000 people placed preorders before the first official sale day (Ramstad 2009). Once it was released, the sales of iPhones were dramatic. More than 100,000 handsets were sold in the first 10 days, and two million were sold in about a year. The introduction of the iPhone triggered the rapid growth of smartphones in Korea, as elsewhere. The change from mobiles as feature phones to smartphones is undeniably unique. The number of smartphone users spiked, again, consisting of 74.1% of total mobile phone users as of December 2015 (Ministry of Science, ICT and Future Planning 2016). In addition, the use of apps, as explained in later chapters, soared, and Korea all of a sudden became a smartland where everybody enjoys cutting-edge smartphone technologies, including apps, and participates in smartphone cultures.

The iPhone effect created a chain reaction as well. The country's biggest seller of phones, Samsung Electronics, had to cut the price of its most advanced and expensive phone, a touch-screen model called Omnia2 (Ramstad 2009). Korean firms have grown under the protection of high trade barriers, which have helped Samsung Electronics and LG Electronics become the world's second and third largest handset makers.[6] But local customers paid the highest prices in the world for mobile phones and among the highest for wireless service, which resulted in the growth of their market share (C. K. Park 2009).

In fact, the explosion of the iPhone in Korea needed to change the domestic wireless industry by putting pressure on competing wireless carriers and handset manufacturers. The tremendous success of iPhones both globally and nationally consequently awakened the domestic handset manufacturers, such as Samsung Electronics and LG Electronics, to the realization of, again, what they missed in the market and what they should do for the future (Korea Telecom 2010). The impact of the iPhone on Korean society has been called the "iPhone shock" because "it influenced not only the industrial practices of both Korean telecommunication companies and local handset makers, but also mobile phone use" (D. Lee 2012, 63): "The delayed introduction of iPhone to the Korean market and subsequent 'smart phone shock' that rippled through the nation's ICT sector offer a powerful illustration of the challenge posed for Korea by digital convergence" (Oh and Larson 2011, 106–8). As will be further discussed in chapter 4, Korea was the 88th country in the world to introduce the iPhone. Apple's top-selling iPhone made its debut in Korea with experts saying that it would likely serve "as a wake-up call for an IT industry basking in an isolated market" (*Independent* 2009). The debut of iPhone in the country brought about a paradigm shift toward mobile applications and contents. The iPhone's philosophy is centered on applications and contents, and mobile telecommunication is just a part of its software (C. K. Park 2009).

The KCC indeed interprets the emergence of the post-iPhone smartphone as "a stepping stone toward the second Internet revolution" (2010, 6, cited in P. Kim 2011). The KCC analyzes why Korea, a strong IT nation, overlooked the oncoming development of the smartphone and its impact. The iPhone was not just another smartphone introduced to Korea. It sparked enormous social debate, ranging from critiques of the Korean government and IT companies to analyses of deeper structural problems in Korea; it truly shook up the country (Ramstad 2009, cited in G. Kim 2011).

As Oh Myung and James Larson (2011, 106) aptly put it, "Despite the extremely rapid diffusion of CDMA-based digital mobile telephony in Korea, the nation was ironically one of the slower ones in the world to actually start using mobile broadband services. The mobile broadband era in Korea only took off after Apple's iPhone arrived there fully two and a half years after its launch in the U.S." "Though its debut was belated, the pace of the iPhone's penetration into Korea's tech-savvy market was about twice as fast as that of other overseas markets that adopted the phone earlier" (*Yonhap News* 2010). The arrival of the powerful handheld computers now referred to as smartphones caught Korea's government and its industry, both handset manufacturers and mobile service providers, off guard (Hopfner 2010). "The change involved more than just a new handset, but a shift toward a smartphone based software ecosystem in which applications, or 'apps' would rapidly proliferate" (Larson and Park 2014, 357). Since the arrival of the iPhone, Korea has rapidly advanced smartphone technologies to become a smartland Korea in which smartphones take a major role in the ICT-driven economy and culture.

Mostly in the 2010s, the primary role of the government has shifted from the protection of the domestic mobile market to the development of new strategies in the smart era. In January 2014, the Ministry of Science, ICT and Future Planning, a newly established communication technology ministry, announced its plan to inject $1.49 billion through the year 2020 into local firms to build the fifth-generation network (5G) in Korea. The 5G network is expected to be 1,000 times faster than the existing long-term evolution (LTE) service, allowing users to download an 800-megabyte movie in one second, compared with 40 seconds on the LTE-Advanced network, currently billed as the world's fastest. The ministry said it had been in talks with the country's three mobile carriers as well as tech firms including Samsung and LG since May 2013 to draw up a plan aimed at taking the lead in mobile network technology. Under a blueprint dubbed "Creative 5G Mobile Strategy," Korea would first develop key features for the new network, which include ultra-HD and hologram transmission as well as a cutting-edge social networking services (SNS) (Ministry of Science, ICT and Future Planning 2014b).

Until the pre-smartphone era, the government supported the mobile telecommunications industries through its protective measures; however, again, it developed its long-term plan to initiate the growth of the era in which smartphones would take a leading role. The current Park Geun-hye (Pak Kŭn-hye) government is conservative, and it is expected to develop a neoliberal economic policy. Contrary to common expectations, the Park government has also initiated the growth of ICTs as the primary engine for the creative economy the government is pursuing. Whether conservative or liberal, the Korean government has continued to develop protective—instead of fully open—and future-oriented telecommunications policies. The current Park government is developing a long-term plan because, as a daughter of Park Chung Hee, who once dictated and developed a state-driven, top-town economic policy, Park Geun-hye knows the effectiveness of a government-driven, long-term economic policy more than anyone else. Unlike the 1970s—the developmental era driven by heavy industry, the current government plans to advance the national economy with ICTs, including telecommunications; we can expect to see more developmental, not neoliberal, telecommunications policies in the coming years. The Korean government has driven the state-led telecommunications transformation, and it has set up concrete partnerships with transnational corporations through a complex negotiation process.

Conclusion

Korean mobile telecommunications industries, both service providers and handset manufacturers, have fundamentally shifted since the 1990s under neoliberal globalization. Domestic politics and economic situations, as well as global forces, have been crucial in the process of reorganizing the Korean mobile telecommunications market. Constant attempts by chaebol to dominate the mobile telecommunications business and changing telecommunications policies as a result of negotiations between international players and domestic actors have played significant roles in the growth of Korea's mobile telecommunications industries. Conglomerates, including Samsung Electronics and LG Electronics, have become the main drivers of Korea's ICT-based economic growth. For Korea, the increasingly mobile digital network environment highlights its overwhelming dependence on hardware manufacturing and exports, in which the chaebol have been dominant, and the urgent need for a transition to greater emphasis on software, content, and services. "Such a transition would increase the relative number and influence of viable small and medium sized enterprises in the nation's economy. Advances in manufacturing technology and the aggressive entry of China into this field with thin, powerful smartphones, only accentuates the challenge facing Korea" (Larson and Park 2014, 357).

In addition, foreign actors, including the U.S. government, international agencies, and TNCs have played pivotal roles in the process of restructuring the Korean mobile telecommunications sector. Their demands to freely compete in the Korean telecommunications market have been strong, and consequently have spurred the government to initiate the process of reshaping the mobile telecommunications industry. Indeed, the strong market-opening pressure from foreign forces propelled the Korean government to change the direction and contents of reform (Jin 2011).

Meanwhile, as a primary player, the government adopted neoliberal restructuring of the mobile telecommunications sector. However, unlike conventional neoliberal wisdom, the Korean government did not immediately ease entrance barriers in the mobile telecommunications sector for foreign-based transnational capitals, opting instead for protection of domestic mobile telecommunications systems. The government has continued to promote the development of mobile telecommunications industries to improve Korea's digital economy. The need for market-conforming methods of state intervention in the telecommunication sector remains an imperative, although the information revolution is changing the very nature and scope of markets.

The Korean government has developed the country's mobile telecommunications industries as a major sector of ICTs in the midst of conflicting relationships with both foreign players and domestic-based TNCs. Regardless of its continuing adaptation of neoliberalism in Korean society, the government, whether conservative or liberal, has continued its role as a forceful state in the ICT sector, including mobile telecommunications, because ICTs are necessary engines for the growth of the digital economy and culture. What we have seen is that physical control by the central government remains far more significant than foreseen by globalist predictions of a retrenchment to small government vis-à-vis Korea's mobile sector. Of course, this does not imply that the Korean government has continued its same level of state power shown in the 1970s and the 1980s; instead, while we should admit the significant role of government-led ICT development, including mobile communications, it is perhaps more accurate to call the Korean approach a "development through a government-industry partnership" (Larson and Park 2014, 349). These two major actors sometimes work together and at other times fall into serious conflicts; however, during the negotiation process, they have developed the Korean ICT-driven industrial and economic structure in the era of neoliberal globalization.

4

Mobile Communication, Globalization, and Technological Hegemony

Kristine Choi (20) has lived in Vancouver, British Columbia, in Canada since 2004. When she moved to Vancouver, she came with her entire family. Her parents and older sister went back to Korea, while she continued her college education. However, she does not feel lonely, because she connects with her family everyday via smartphones and Kakao Talk. Whenever she wants to talk with family, she just taps a telephone number or sends a message, anytime and anywhere. Although she and her family live in different countries, they feel as if they live together. (notes from an interview with a twenty-year-old female student in Vancouver, Canada)

The rapid growth of smartphones has changed the environment in which people communicate with each other because smartphones allow people to connect easily, not only through a telephony system, but also through one of several instant mobile messenger applications. As globalization primarily implies interconnectivity and interdependence among people and countries (Tomlinson 2000; Ritzer 2011), the newly developed smartphones in the 21st century have actualized connectivity much faster and more easily than in previous decades. Previously, "interconnectivity happened through structural and institutional linkages" through mass media, including television systems (Waisbord 2004, 359); however, smartphones have extended global interconnectivity to global citizens who connect with their family and friends anywhere. Smartphones can "contribute to an increase in social interactions among spatially distributed family members and acquaintances in daily life, thereby strengthening the internal group bonds of close relations" (D. Lee 2013, 274).

A new generation of young and tech-savvy consumers, including Kristine herself, uses the smartphone as a digital platform to fulfill their dream of being global citizens. The ability to be constantly connected to the Internet and apps via smartphones has remarkable implications in the era of globalization. It is becoming more evident that there is a blurring of lines between geographical locations because people always connect through smartphones and smartphone apps, such as Twitter and Kakao Talk, as well as other social media. This is possible because smartphones can produce, receive, and send video clips and messages and receive popular culture (e.g., television programs and films), while camera phones are the first form of mobile phone convergence that has introduced images (Fortunati 2012). As Ling (2008, cited in D. Lee 2013) points out, existing mobile phones have allowed family members and friends to enjoy full-time connectedness to and ritual interactions with one another, and consequently, they have helped to reinforce social bonds in close relations. D. Lee (2013, 273) also states, "The frequent exchanges of humor and greetings via text messages could strengthen social bonds, not because of their content but because of their assurance of being connected." However, the smartphone era is beyond what the traditional mobile functions provide, because through Kakao Talk, Line, and Skype, they enjoy not only telephony and text messages but also video chatting anytime and anyplace. Globalization arguably diminishes the relevance of borders and territory, and smartphones seemingly actualize our contemporary assessment of globalization. However, as a reflection of their short history, it is rare to find academic discussions on the role of the smartphone in the process of (neoliberal) globalization.

This chapter addresses the question of globalization in the era of smartphones by analyzing Korea's reception of the iPhone in 2009 and the recent growth of locally made smartphones, including Samsung's Galaxy series, in the global market. This chapter develops new perspectives in the existing body of knowledge on the issue of globalization by discussing its pertinence to smartphones. It discusses Korea's smartphone growth within the socioeconomic structure of the global capital system. It finally advances our discourse on the primary characteristics of digital platforms (Jin 2015), which U.S.-based platform technologies and culture dominate in the global market.

Cultural Globalization versus
Globalization in Technology

While there are several theoretical approaches to globalization, there is an agreement that the increasing interconnectedness between individuals, businesses, and regions affects the lives of people (Giddens 1999; Castells 2000). ICTs play a key role in the globalization process, primarily because of their role as the facilitators of interconnectedness. As Castells (2000) points out, globalization can be fueled by information technology, mainly through "information-processing devices." However, simple interconnectedness alone cannot explain the complexity of globalization (Giddens 1999; Lechner 2009), because the contemporary form of globalization driven by ICTs is based on connectivity, which signifies the power shift of capital (K. S. Lee 2008). "It is crucial to comprehend the new material conditions of globalization, which is virtually constructed by the worldwide electronic network of capital," in this case through smartphones (K. S. Lee 2008, 3).

With the recent growth of platform technologies, many researchers and policymakers have begun to emphasize the flow and use of platform technologies, such as social network sites (e.g., Facebook and Twitter), smartphones (including operating systems), and search engines (e.g., Google), in order to comprehend whether ICTs have changed the contours of global markets (Gillespie 2010; boyd 2011; Jin 2015). As several scholars argue (Pieterse 2006; Boyd-Barrett 2006; Shi 2011), ICTs constitute an important factor in the globalization trend. However, despite the importance of ICTs to the globalization process, scholars have for the most part neglected the political economy of ICTs and their role in reinforcing the U.S. superpower, or their development in non-Western countries that may unsettle U.S. hegemony (Boyd-Barrett 2006; Shi 2011; Boyd-Barrett and Xie 2011). In particular, it is critical to understand whether the smartphone era, which is also driven by applications, has changed the direction of the one-way flow in technologies and relevant systems from Western countries, primarily the U.S., to non-Western countries because of the emergence of local forces, in terms of both smartphone devices and apps.

To historicize the recent growth of the smartphone and its implications in the process of globalization, one first needs to address the process of globalization in popular culture, because a comparison between popular culture and technology, in this case, smartphones, provides us with a reliable interpretation for comprehending how the mobile industry has become a major force in the globalization process. As Daya Thussu (2006; 2007) clearly points out, in the realm of popular culture, the globalization process mainly goes through four stages: (1) the spread of the U.S. model of professional commercial culture that has brought changes in the national media industries, (2) the production of conflict between Western (primarily American) culture and local culture, (3) the cultural revival in several non-Western countries as a result of this discontent, and (4) finally the export of local popular culture to neighboring countries and then potentially to Western countries.

This process normally applies to almost all cultural products, such as television dramas, film, and music. For example, American film producers began to penetrate the global market as early as 1910s and 1920s, and commercial broadcasters started their global penetration in the 1960s and 1970s, which provided empirical grounds for cultural imperialism theory (Gomery 1996; H. Schiller 1976), and since the end of the Cold War, global television has helped promote Western consumerist culture in non-Western countries. While some aspects of globalization brought beneficial changes in national media industries because they strengthened liberal democratic culture, Western influence has produced discontent in many countries because of the sexual and violent content in Western television programs (Thussu 2006). Several countries have vehemently resisted Western cultural products in the name of the protection of local culture and cultural identity, as shown in the screen quota debates in Korea, Mexico, and France. This does not mean that they deny the inevitable homogenization of everything occurring due to the strong influence of American culture. Instead, it means that at least a few non-Western countries protect the way they are, or want to be (Goldsmith and Wu 2006).

Consequently, many countries have experienced a sudden growth in local culture, as in the case of telenovelas in Mexico and Brazil, as well as Korean television programs and films. In other words, the reaction to Western television programs has been the growing popularity of these countries' television dramas, which has resulted in the reversal of cultural flow, from non-Western to Western countries. Several theoreticians (Curran and Park 2000; Iwabuchi 2010; Waisbord and Mellado 2014) argue that non-Western countries have slowly developed their own unique local cultures and exported them to other countries, and they dub this 'de-westernization.' The process, of course, takes at least several decades, and, in fact, only a few countries are able to develop their own cultural power. Since technology is not separated from culture, and the process of globalization in the field of new media is not much different from that in popular culture, it is useful to critically employ this understanding of the process while examining the realm of smartphones.

The Globalization Process in the Smartphone Era:
The iPhone goes Local

In the 21st century, digital platforms, in particular smartphones, have widely been viewed as the essential catalyst of contemporary globalization. As the smartphone becomes more pervasive and as more and more aspects of life become digitalized, "It is becoming much easier for human beings everywhere to access, learn from, share, and improve upon the impossibly varied and plentiful information available on the phone" (Goldsmith and Wu 2006, 179–80).

As discussed in chapter 2, the history of smartphones is relatively short, and technically speaking, the successful evolution of iPhone as a commercial smartphone occurred less than 15 years after IBM's Simon was invented in the early 1990s. Apple developed and started to sell its commercial smartphone in 2007. As in several cultural industries, the spread of the U.S. commercial culture in the form of the smartphone into other countries has been much expected, as Apple was the first transnational telecommunications giant that developed commercially successful smartphones, but it was even stronger than the normal flow of cultural products, achieving sudden global penetration within only a few years of the iPhone's introduction in the United States.

As Thussu (2007) points out, the United States leads the global export market in media products; however, when it comes to smartphone technologies, the magnitude is even larger and stronger than with popular culture. Apple's iPhone has rapidly penetrated the global smartphone markets and seemingly confirmed the leading role of the United States in the mobile sector. After other successful platforms, including Facebook, Google, and Twitter, the United States has extended its dominant position with Apple's iPhone. This new phenomenon contributes to the imperialist incursion of globally dominant technological cultures in local contexts. In other words, the phenomenal growth of Apple's iPhone in the global markets has provided a promising case study to confirm whether the United States and a few Western countries dominate or if non-Western countries can change the contours of the direction of the flow in technology and mobile culture. This is significant because globalization based on "mobility and connectivity signifies a power shift of capital" and "forces local states to affiliate with or integrate into a part of the new world system." The survival of local regions indeed depends largely on "their close links to the global electronic conduits of capital" (K. S. Lee 2008, 3).

As is well chronicled, Apple has swiftly penetrated global smartphone markets, including Asia. "The iPhone and its global distribution symbolizes Apple's strategies as a global transnational corporation to control not only the market of the smartphone itself but also its software development environment, wireless services, and the information and entertainment available to its users" (Shi 2011, 134–35). What is interesting are Apple's globalization strategies, as will be detailed in the following section. Once Apple launched its iPhone in 2007, the first-generation iPhone was sold only in Germany, the UK, and France in 2007, and Ireland and Austria in the spring of 2008. The second generation rolled out across the globe in 22 countries and regions in July 2008, including Australia, Canada, Denmark, Egypt, Finland, and Hong Kong (Apple 2007a; Apple 2007b; Shi 2011). Other countries were anxiously waiting iPhones, which consequently decreased negativity in terms of the hegemonic power of the U.S.-made smartphone in many countries.

Shipments of the smartphone more than doubled in the first quarter of 2010, and Apple attributed most of the growth to demand from Asia and Europe. Apple highlighted "incredible demand" for the device in newer markets like China and Korea. As iPhone sales exploded overseas, apps also started to go global (Kane and Worthen 2010). When Apple started to sell iPhones in Asia, many Asians sought out this cutting-edge, 21st-century device. In China, for example, hundreds of people camped outside Apple's or its local carriers' stores just to be among the first to own this gadget.

> [O]bservers wondered, "What are these people really buying? The phone? Or something it symbolizes that is driving this 'iWant it and iWant it now' attitude?" The editorial of *South China Morning Post* (2008) explains the phone's cultural influence very well: "The iPhone, clearly, is no mere electronic accessory, but a force of 21st century globalized culture . . . if a person doesn't appreciate it, he or she just doesn't get it and therefore deserves to be consigned to the dustbin of cultural history." (Shi 2011, 138)

Of course, the global penetration of the iPhone was followed by other goods, sold in the Apple Store. "That iPhone developers are pursuing international strategies is a sign of how Apple's App Store business—which offers games and productivity tools that can be downloaded onto the iPhone, iPad, and iPod touch—is starting to mature, with some developers now having established enough domestic traction that they can turn their sights to broader opportunities" (Kane and Worthen 2010). Apple's iPhone has successfully achieved its goal of global penetration, and Apple, the U.S.-based transnational corporation, reigns supreme in the realm of smartphones and apps.

Rising Local Resistance, but Not in the Realm of the iPhone

What is unique about the spread of the iPhone in the global markets is a conflicting mix in the process of adaptation of this new gadget. As several cultural products, such as film, music, and broadcasting, prove, the production of disconnects between Western culture and local culture has occurred in many countries. In the realm of popular culture, for example, several countries, including France, Canada, and Korea, have developed both screen quota systems for the national film industries and program quota systems for the national broadcasting industries (Thussu 2007; Jin 2011; Brownell 2015). Many Middle Eastern countries and Asian countries, including China and India, have also tried to block Western popular culture in the name of the protection of cultural identity and sovereignty. Non-Western discontent with Western countries remains strong in the popular culture sector.

The smartphone experienced two different receptions in the Korean market. On the one hand, the government and telecommunications manufacturers had a strong interest in delaying the import of iPhones in order to develop domestic smartphones. The iPhone encountered various obstacles in several countries, including Korea and China. As usual in these Asian countries, some of the obstacles are generic to most media globalization efforts in Korea, namely, state regulation and conflict of interest with domestic companies (Shi 2011).

On the other hand, unlike in the realm of culture, there are no serious conflicts between American iPhone culture and local mobile culture among users. Apple's unique globalization strategy partially worked here. Apple did not rush, and Apple was not much interested in penetration of the Korean market because of its small size. As I have briefly discussed, Apple planned to monopolize the entire value chain of the iPhone (hardware, software, content, and service); therefore, the company carefully selected its global partners.[1] As Apple first partnered with Cingular (owned by AT&T Inc.) as the sole service provider for iPhones in the United States, Apple similarly partnered with only one service provider in many countries (Apple 2007c). Apple's strategy delayed its launch of iPhones in several countries, which did not create any serious conflicts with local users.

Against this backdrop, the market situation in many countries was rather interesting, because many Asians insistently asked their governments to import the iPhone, while the governments and the industries delayed the importation. Although there have been some tensions between the U.S. and other countries, many citizens and user groups in these countries, including China and Korea, acted differently toward the smartphone. Potential consumers in these countries developed their own strategies to use iPhones, while demanding the government liberalize the home market. For example, in the Chinese market, "Chinese cell phone users and a virtual community of 326,000 Chinese iPhone fans called 'bbs.iphone' actively created ways to participate in the global iPhone phenomenon with smuggled units. The users' consumption culture and their activities of buying smuggled iPhones, unlocking them from the chosen carrier, and jailbreaking the official software system worked to demand Apple and the Chinese government . . . release the iPhone in China" (Shi 2011, 143).

In Korea, Apple also had a large preexisting fan base because it had an innovative and "cool" cultural image. A long queue of fans waiting on the iPhone's new model on launch day was an emblem of cool in popular culture and reflected the high degree of brand loyalty among Apple fans. Korea's largest iPhone user community (http://cafe.naver.com/appleiphone, 1.27 million members, as of July 2012), known as *Asamo* ("Apple lover") in Korean, has been a venue for knowledge sharing, opinions, and collective purchases related to the iPhone since December 2006, three years before the Korean introduction of the iPhone. Users in this community exhibited critical attitudes toward the WIPI policy, and called for the abolition of the WIPI mandate by means of an online petition (Bae 2014, 143). These potential iPhone users, instead of resisting this Western-made smartphone, played a significant role in admitting Apple's iPhone into the country because of their high expectations for this cutting-edge technology, which is very different from the case of popular culture. When Apple released iPhone 6 and iPhone 6 Plus in October 2014, many Koreans again expressed their strong desire to get iPhones, with no serious resistance to the products. It is estimated that in Korea there are three million Apple maniacs (C. Kim 2014). From a critical cultural perspective, the market success of a consumer product relies on a cultural pattern (Williams 1999, cited in Shi 2011), whereby the product is invested with social and personal meanings:

> In the circulation process, the meaning of iPhones as products of human labor—sometimes under unfair employment conditions—is stripped away, which "leaves an empty symbolic space" quickly filled with new meanings such as youth, status, coolness, freedom, unconventionality, technological and fashion-forwardness, and, in a global setting, American-ness, meanings that are more suitable to consumption and capable of explaining the consumer mania described above. Overall, with the iPhone, Apple presents to the world a whole package of corporate globalization. (Shi 2011, 138)

Moreover, members of the iPhone community are extremely proactive by staying in sync with the global iPhone phenomenon (Shi 2011). Smartphone culture is also distinctive in many different countries because of the various forms of convergence between hardware and software (e.g., smartphone and Kakao Talk); however, it did not take much time for Apple to penetrate the global markets with no particular resistance, as opposed to popular culture.

When Koreans started to import iPhones and Samsung and LG also began to sell their smartphones in 2009, many Koreans preferred iPhones to Samsung's Galaxy with no hesitation. Apple landed in Korea smoothly, unlike some other parts of American popular culture, a landing that was effective in that it did not cause any serious anti-American mentalities.

Technological Breakthroughs in Local Industries

Technological revival, which is the third stage of globalization, in several non-Western countries, such as Korea, Taiwan, and China, is not only the result of discontent (primarily among the government and telecommunications corporations), but also the result of active participation and/or demand from iPhone fans. This means that local smartphone manufacturers in these countries are able to develop locally created smartphones. In other words, the technological breakthroughs in several non-Western countries in the mobile communication sector occurred not because of a result of discontent, but because of both the political protection of the local mobile industry and local fans who rapidly consume cutting-edge smartphones.

As in the realm of digital technologies, non-Western countries are not just consuming Western-made iPhones. Several of these countries' transnational corporations have swiftly developed their own smartphones and increased their presence in the global markets, competing with iPhones. "With digitalized information, ICT corporations and traditional media conglomerates in developed countries seek to harness a variety of information platforms and reach different audience segments" (Shi 2011, 134). In fact, within only a few years, Samsung has become the global frontrunner in the realm of smartphones, based on its previous mobile technologies.

Under favorable government policies, as discussed in chapter 3, several domestic handset makers, such as Samsung and LG, and a few wireless telecommunications service providers have advanced their devices and services, resulting in a smartphone revolution. While mobile communication technologies are not new for these transnational corporations, their move to the realm of smartphones has been a big breakthrough. Korea has advanced several significant digital technologies, and smartphones and related apps are amongst the fastest and most advanced. For example, it took 86 years for traditional fixed-line phone service to reach 10 million domestic subscribers. It took 30 years for television to do the same, and the nation's personal computer penetration reached 10 million 16 years following its introduction, while mobile phone subscribers reached 10 million 14 years after the first service in 1998. For the Internet, it took 13 years.[2] The smartphone era has been unprecedented. It took only 16 months to reach 10 million subscribers, while for Kakao Talk, it was only 13 months (*Munhwa Ilbo* 1999; D. Kim 2002; Ministry of Science, ICT, and Future Planning 2014a) (fig. 4). This suggests that the time period to reach 10 million subscribers is getting shorter for new technologies, and the smartphone era proves the revolutionary growth of use, symbolizing the significance of smartphones and apps for their users.

Road to the Final Stage: Local Goes Global

The rapid growth of smartphones has changed the road of Korea's globalization in the mobile communication sector. Smartphones and relevant applications have taken primary roles in the networked society, and the globalization process cannot be the same as that of the 1990s and the early years of the 21st century, when global forces primarily took a major role. Local forces have developed primarily because of smartphones' enormous potential, not only for capital accumulation, but also for sociocultural op-

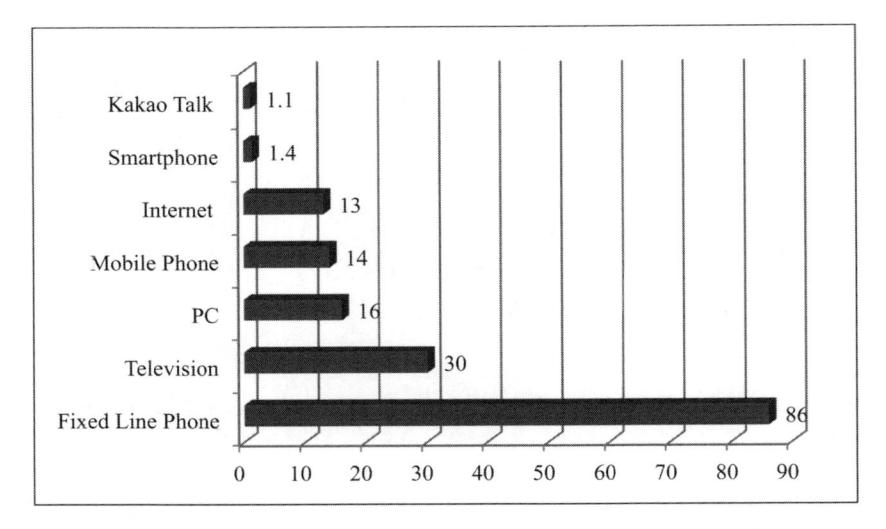

Fig. 4. Time to Reach 10 Million Subscribers (unit: year). (*Source:* Data from Munhwa Ilbo 1999; D. Kim 2002; Ministry of Science, ICT and Future Planning [Mirae Ch'angjo Kwahak Pu] 2014a; Yi Hyŏng-gyŏng 2014.)

portunities. Based on their phenomenal growth in the domestic market, Korean smartphone makers have begun to penetrate global smartphone markets. Korea has continued to increase its exports of telecommunication goods and services to the world, and Korea has, once again, advanced mobile technologies, first with feature phones, and then with smartphones, and it has played an important role in the global markets.

This final stage of the globalization process in the smartphone sector is not the same as we have seen with popular culture, and the smartphone has played an important role in the global markets. In the case of popular culture, several emerging markets have begun to export their cultural products to neighboring countries and then gradually to Western countries, as Mexico's telenovelas and the Korean wave—the sudden growth of local cultural industries and their exports of domestic cultural products—exemplify. In popular culture, Mexico, Brazil, and Korea have penetrated neighboring countries primarily based on either "cultural proximity" (Straubaar 1991) or "differences" (Jin and Lee 2012) and then exported cultural products to Western countries, although they are still marginal in North America and Western Europe. In the realm of smartphone, Korea has not only targeted Asian countries, but has also aimed at North America and Europe almost at the same time.

More specifically, two different stages have emerged. When Korea exported feature phones, it was not significant in the global mobile market. However, since 2009, Korea has increased its export of smartphones (Ministry of Knowledge Economy 2013), which changes the map of the global smartphone industry. When Nokia was number one in the worldwide mobile device market in 2009, it held a 38.9% market share. Samsung held the fifth position (3.3%) (International Data Corporation 2010). Only three years later, Samsung became the number one smartphone handset maker and exporter, with 23.4% of the market (*Yonhap News* 2013).

As table 1 shows, Samsung has extended its lead as the world's biggest smartphone maker, accounting for 32.5% of sales, ahead of Apple's 16.6% in 2013 (Trendforce 2015), although it went down to 28% in 2014, and again to 24.1% in the first quarter of 2016. While Samsung's share has recently decreased, its leading position in the global mobile markets continues (Communities Dominate Brands 2015; International Data Corporation 2016). In fact, based on the survey conducted by International Data Corporation, *AFP* (2014) reported that total smartphone sales reached 494.4 million units worldwide in 2011, doubling that volume in just two years in 2013, and Samsung saw growth of 42.9%. The U.S. has become one of the largest markets for Korean smartphone manufacturers. The Western markets, including the U.S. and Western Europe, are much more significant for local smartphone makers than for corporations in other Korean cultural industries (Jin 2016).

Korean smartphone makers, again, have successfully appeared in these markets. As a result, the smartphone has become Korea's first major ICT that has penetrated global markets as well as Asian markets on a large scale and at the same time within such a short period of time. Some policymakers and IT experts claim that Korea has changed the contours of globalization, from Western-dominant to non-Western-driven interconnectedness. In particular, they might argue that the case of the smartphone resolves unequal power struggles through the emergence of the local technology industry.

The global smartphone markets will become more competitive, because Apple in the U.S. and a handful of locally based smartphone makers in China and Taiwan are intensifying their global presence. For example, China's Lenovo acquired Motorola's handset division in January 2014 for $2.9 billion. The purchase gave Lenovo "a beachhead to compete against Apple and Samsung as well as increasingly aggressive Chinese smartphone makers in the highly lucrative U.S. arena" (Damouni et al. 2014). Huawei and Xiaomi in China have also extended their role to become major players. The global handset markets are the front lines for both Western and non-Western countries. In the near future, Korea, the U.S., and China will be competing to expand their positions in the global markets, which may consist of a three-way hegemonic dominance in the realm of hardware. This means that it is possible to witness the growing role of non-Western countries, although the U.S. will not give up its strong position in handset markets.

How to Understand Globalization in the Smartphone Era

Several key factors contribute to the global penetration of both popular culture and technology; one of the common drivers of such globalization is that transnational smartphone makers, including Apple and Samsung, have developed localization strategies as a form of outsourcing. Since globalization in economic terms refers to "the growing economic integration of the world, global trade and investment increasingly cross international borders." During the process, "the spread of ICTs is changing the way companies organize production, and increasingly allowing services as well as manufacturing to be globalized" through localization (Schifferes 2007).

Accordingly, localization is a well-known scheme for transnational corporations to penetrate the global markets. In the realm of media and communication, localization is a strategy that global media agencies need

TABLE 1. Top 10 Smartphone Makers

Rank	2013	Market Share	2014	Market Share	2015 3Q	Market Share
1	Samsung	32.5	Samsung	28	Samsung	23.7
2	Apple	16.6	Apple	16.4	Apple	13.5
3	Lenovo	4.9	Lenovo/ Motorola	7.9	Huawei	7.7
4	Huawei	4.4	LG	6	Lenovo	5.3
5	LG	4.3	Huawei	5.9	Xiaomi	5.2
6	Sony	4.1	Xiaomi	5.2	LG	4.2
7	Coolpad	3.6	Coolpad	4.2	Nokia	4
8	ZTE	3.2	Sony	3.9	Oppo	3.5
9	Nokia	3	ZTE	3.1	Vivo	3.4
10	Rim	2.5	TCL	2.7	TCL/ Alcatel	2.9
	others	20.9	*others*	15.8	*others*	26.4

Source: Data from TrendForce 2015; Communities Dominate Brands 2015.

to develop to overcome the barriers that cultural differences pose to global flows of culture. Localization refers to "a range of institutional and textual arrangements that rework foreign content to be more culturally relevant for domestic viewers; dubbing and subtitling are forms of localization, as are rewriting and re-recording popular songs in the local language with local pop stars. When localization strategies apply to the hardware sector, it appears in several forms, including the establishment of local branches of global media firms, as in the case of MTV, and/or outsourcing" (Havens and Lotz 2012, 238–40).[3] The lure of outsourcing is primarily the cost savings of hiring cheaper production workers. Outsourced production has become more feasible because "the global communications infrastructure makes communication with foreign labor cheap and relatively quick while also permitting domestic supervisors to maintain a good deal of control over foreign workers" (Havens and Lotz 2012, 240).

Apple, as a U.S.-based transnational corporation has exploited cheap labor in China, which is a typical form of value exchange—bringing large profits to the U.S. In fact, for years, nearly all of the world's iPhones and iPads rolled off the assembly lines of a single company—Foxconn, a transnational electronics contract manufacturing company in Taiwan but with several factories in China. Apple designs the iPhone but contracts with nine different companies from four different countries (Japan, Korea, Germany, and the U.S.) for the required components. These firms ship their products to China, where the actual assembly of all modular components into finished iPhones is done (Oh and Larson 2011), initially by Foxconn, which oversaw their assembly and export to the U.S. and the rest of the world (Hart-Langsberg 2013). Apple has utilized the division of international labor, as Miller and Leger (2001) argue, in order to benefit from the cheaper labor cost in China.

However, Apple has gradually diversified its outsourcing strategy to avoid potential risk, moving assembly from Foxconn to Pegatron, another Taiwanese electronics-manufacturing company. Foxconn's cost advantages from scale have waned as it works to improve factory conditions after a spate of high-profile worker suicides and accidents in recent years (Qiu 2009; Dou 2013). Although briefly caught in the public eye in 2011 because of a factory explosion that injured dozens of workers, the company has largely escaped the spotlight that initially forced it to increase wages and make changes to its labor practices. "Foxconn, in its growing heft as the world's largest electronics contract company, was also getting more difficult for Apple to control, with incidents such as the changing of component sourcing without notifying Apple" (Dou 2013). At the same time, "Foxconn became frustrated with the growing complexity of Apple products, such as the iPhone 5, which is difficult to make in the volumes Apple needed." Therefore, under current chief executive Tim Cook, Apple is dividing its weight more equally, using a relatively unknown supplier, giving the technology giant a greater supply-chain balance (Dou 2013). The production of Apple products highlights why core transnational corporations embrace cross-border production.

Like Apple—a Western-based transnational corporation—Samsung has conducted a globalization strategy to accumulate capital. The Korean smartphone maker has strategically developed its localization process, establishing several regional plants. When it launched a cell phone factory worth nearly US$700 million in northern Vietnam in 2009, Samsung already had six mobile phone manufacturers in other countries, including China, India, and Brazil (Hung 2009). Through both localization and outsourcing strategies, Samsung has utilized a globalization logic in the field of the smartphone. "Samsung built the world's largest smartphone business by tapping into China's cheap and abundant workforce." However, it has shifted its output to Vietnam "to secure even lower wages and defend profit margins as growth in sales of high-end handsets slows" (Lee and Folkmanis 2013). As Martin Hart-Landsberg (2013, 13) points out,

> Capitalism is not a static system. The levers driving its motion are capital accumulation, competition, and class struggle. Their complex interplay generates pressures and contradictions that compel profit-seeking capitalists to continually reorganize their activities, a process that has profound consequences for our lives. In other words, our social condition is largely shaped by the actions of the leading business organizations. Today, these business organizations are transnational corporations. Their drive for profit has produced a new, more globalized stage of world capitalism.

While recognizing the remarkable penetration of Korean smartphones in the global markets, we need to ask whether technological penetration from non-Western to Western countries, known as *counterflow*, has changed the asymmetrical power relations between the global and the local. Given the global penetration of Samsung and LG smartphones, some researchers and policymakers consider that the uneven flow in the realm of smartphones, and digital technologies overall, has disappeared. They recognize that the successes of Korean smartphone makers, at least in recent years, in the global market, including North America, are a new trend. As discussed, the U.S. initiated the smartphone revolution in the mid-1990s, and Apple commercialized the first smartphone in the world globally. However, the smartphone is no longer an American phenomenon. Korean smartphone makers have jumped on the bandwagon, and these handset manufacturers have swiftly become global leaders in the smartphone sector. Korean hardware is now being used globally, and Korea has played a major role in the development of digital technologies (Jin 2015, 162). Once peripheral, Korea has achieved the status of digital empire in the smartphone era.

Of course, while Korea has achieved a global presence with smartphones, the future for Korean smartphones may not be rosy. There are two urgent issues for the Korean mobile communication industries and policymakers: one is the recent setback Korean smartphones have experienced in the market, and the other is the increasing role of U.S.-based operating systems. The very recent smartphone market is not healthy for Korean smartphone makers. Samsung and LG lost global market share in 2014 and 2015 amid severe competition with Apple's iPhone and Chinese smartphone makers. China in particular is one of the markets where Samsung is facing tough competition, as local brands, including Xiaomi, the world's third largest smartphone maker—only behind Samsung and Apple—Huawei and Lenovo, have been aggressively expanding not only on their home turf, but also in overseas markets (International Data Corporation 2014b; Rabouin 2015; Goldman 2015). Samsung was still the leader in the global mobile market as of March 2016 (International Data Corporation 2016); however, its position is not as strong as in previous years. Only a few years ago, telecommunication experts expected that the role of Samsung to increase as smartphones surpassed feature phones as a share of the market; however, because of increasing competition from other local handset makers, Samsung's leading role is in jeopardy. More importantly, unlike the hardware side, the U.S. has continued to dominate the operating system side, as discussed in the following subsection. Since U.S. corporations are able to control the smartphone market with their operating systems, local-based smartphone makers in Korea have been able to achieve only limited success in globalization.

American Ways in the Smartphone Era

Mobile media can "loosely be considered software, applications, or services accessed through mobile devices" (Humphreys 2013, 21). Thus, it is necessary to address whether locally based transnational smartphone makers have developed operating systems, comparable to hardware. Unlike smartphone devices, the software side proves that American dominance in the smartphone industry remains and may even be intensifying, because of underdeveloped operating systems from non-Western smartphone corporations. Apple and Samsung have viciously competed in the global markets in recent years; however, primarily U.S.-based ICT firms have benefited from global markets with their own operating systems. The two American-based operating systems, Google's Android and Apple's iOS, are dominant in the global markets. Android and iOS seem to be everywhere, forming hegemonic power in the smartphone industry (Jin 2015). It's safe to call the result a duopoly (Fingas 2013). Samsung's reliance on Android "unquestionably accelerated its growth in handset sales by offering it a 'turnkey' mobile ecosystem. But Android could also turn out to be its Achilles' heel" (Lev-Ram 2013). While Samsung builds some services on top of the operating system and tries to give its Galaxy phones their own look and feel, the company ultimately does not own Android. In fact, the operating system is freely available to all other handset makers, including "up-and-coming Chinese manufacturers that are developing cheaper phones" (Lev-Ram 2013).

Consequently, only two American-based platform giants have fully capitalized on the Korean market, because they benefit from their dominant positions. According to Distimo (2013), a consulting company specializing in the app market, in 2013 the top five countries in terms of mobile app revenue from the Apple App Store and Google Play were, in order, the U.S., Japan, Korea, the UK, and China. The three countries with the greatest market growth in 2013 compared to 2012 were Korea (759%), Japan (280%), and China (240%). This implies that Google and Apple have rapidly increased their app revenues in Korea through the increasing use of Google Play and the App Store by Korean smartphone users. Korea has become one of the most significant app economies for U.S. corporations, as will be detailed in chapter 5. In December 2013, Google indeed notified domestic telecommunications service providers that it changed service fees on Google Play that Google and service providers split from 1:9 (Google's share:service provider's share) to 5:5 (Ju 2014). Based on its monopolistic status, Google has begun to fully capitalize the app market in Korea. Although Android itself is open source, Google has utilized this open source to make revenue through both intellectual property rights and Google Play, a very familiar strategy in monopoly capitalism.

The digital convergence between American operating systems (e.g., Android and iOS) and Korean handsets has brought about noteworthy consequences in several fields. The main issue is that the U.S. has substantially expanded its influence in digitally driven technologies and culture due to platforms, as in the case of popular culture, such as films and music. It cannot be denied that American operating systems are dominating the global smartphone markets, which shows an asymmetrical power relationship that local actors currently cannot overcome. As American cultural products are still influential in the global cultural markets, the U.S. has continued its dominance through control over software in the digital era (Jin 2015).

As Kwang Suk Lee (2008, 6) points out, "Critiques of Marxist economic reductionism, which had a tendency to focus on a shift toward a single united world economy, have arisen first from the problematic that globalization can no longer be understood by simple center-periphery models," and this view also partially reflects the current milieu surrounding the smartphone industries in hardware. However, when we go to software, which is much more significant than hardware, because of Western monopoly, we still witness a hegemonic dominance and an increasing digital divide in the realm of smartphones. Again, the current form of globalization through smartphones signifies a power shift of capital, and forces local states to affiliate to or integrate into the global capital system (K. S. Lee 2008, 3). This implies that an asymmetrical relationship of interdependence in smartphone technologies in both devices and apps between the West, primarily the U.S., and developing countries has increased. The disparities consequently encourage the concentration of capital in the hands of a few U.S.-based platform owners, resulting in the expansion of the global divide, which the following chapters analyze.

In other words, as I have discussed elsewhere, the newly constructed disparity, as a form of platform imperialism, needs to be understood as an unequal power relation among countries backed by software, which strongly favors the U.S. government and transnational firms because commercial values are embedded in smartphone handsets and apps (Jin 2015). As Harvey's notion of the new imperialism (2007) explains, the platform imperialism that has been evolving in the early 21st century entails construction under the hegemony of the U.S. The connection between the rise of platform imperialism within the struggle of neoliberal globalization, as in the case of post-9/11 U.S. politics, is vitally important in understanding the contemporary global order, in which smartphones are highly involved (Jin 2015, 67).

The world is not flat, contrary to what Thomas Friedman (2005) argues. He points out that the development of the Internet browser was akin to the end of the Cold War or the outsourcing of labor from developed countries to developing countries, a flattening force that enables goods and content to move more fluidly across national borders (cited in Jenkins at al. 2013). Friedman (2005, 92) argues that "the spread of the Internet and the coming to life of the Web, thanks to the browser and fiber optics, enabled more people than ever to be connected and to share their digital content with more other people for less money than any time before." In other words, he claims that the consequence of globalization has flattened the world so that the disparity between Western and non-Western countries is reduced. Again, contrary to what Friedman claims, however, the world is not flat, because "there are divides between countries which ensure that not everyone gets to participate on the empowered terms" (Jenkins et al. 2013, 284–86). Globalization through cutting-edge technologies, both the Internet and smartphones, has increased global connectivity, yet the continuing dominance of the U.S. due to its strong footholds in apps on smartphones makes it difficult for people and corporations in non-Western countries to benefit from the emerging app economy. As George Ritzer (2011, 149) aptly puts it, "Global media are dominated by a small number of large corporations [mainly in Western countries], and this is being extended from old media to new media." Smartphones have actualized a flat-world ideology more theoretically than practically.

Indeed, critics of Apple's App Store are concerned that "Apple holds too much power in its relations with independent producers of mobile applications. Apple curates what is made available in the App Store based on what aligns with its own perceived market interests" (Jenkins et al. 2013, 245). As Zittrain (2009) argues, the restrictions of Apple's App Store mean only a select group who agree to Apple's terms can access the tools to create content for its platforms; innovation is concentrated and filtered, if not ultimately regulated, by Apple itself (cited in Jenkins et al. 2013). Zittrain (2009, 19) indeed points out:

> To those who managed to tinker with the code to enable the iPhone to support more or different applications, Apple threatened (and then delivered on the threat) to transform the iPhone into an iBrick. The machine was not to be generative beyond the innovations that Apple (and its exclusive carrier, AT&T) wanted. Whereas the world would innovate for the Apple II, only Apple would innovate for the iPhone.

In other words, the Software Development Kit (SDK) that was released in 2008, and which developers must use to create material for devices such as the iPhone, gives Apple the right to approve the technology, functionality, content, and design of these applications. Only those approved will be sold through the App Store, the only channel through which apps can officially be sold. Apple reserves the right to recall or stop selling apps as it sees fit, to promote certain apps over others, and to prevent the sale of apps that duplicate the functionality of official programs or that provide users with functionality Apple or its network partners dislike (Zittrain 2009, cited in Jenkins et al. 2013, 245–46). Of course, the hardware sector has developed a new look thanks to the increasing role of non-Western-based handset makers, including Samsung; however, as mobile includes both software and hardware, when we consider the software sector, including applications, globalization as a way to advance global quality and evenness cannot be accomplished anytime soon, and the gaps between the U.S. and other countries, in particular, non-Western countries, will worsen in the near future. America's hegemonic position in the ICT sector has been backed by software and relevant intellectual properties, because software means the accumulation of capital in the hands of software designers and owners who are mostly in the U.S. and a few Western countries.

Conclusion

Over the last two decades, information and communication technologies, such as the Internet, social media, and mobile communications, including recent smartphones, have become among the most significant for the global economy and culture. The technological landscape of the smartphone shows an interesting trend, as locally produced digital technologies have become popular in global markets. The globalization process of the smartphone era in Korea went through four stages: (1) the spread of the U.S. model iPhone and thereafter of professional commercial culture in Korea, which has brought changes in national telecommunications industries, (2) the production of incompatibilities between American technologies (in this case, the iPhone) and local technologies (e.g., Samsung and LG) through protective regulatory measures, although users rapidly accepted iPhones, (3) the technological revival of Korea's smartphones as a result of both discontent and acceptance, and (4) finally the export of local smartphone devices to both neighboring countries and Western countries at the same time, but only the devices, not operating systems. Through this process, Korea, as one of the leading countries developing locally produced smartphones, has dramatically shifted the political-economic situation of the country in the ICT sector.

While admitting the remarkable achievement of manufacturing and export of ICT hardware in Korea, one needs to understand that the nation's future rests increasingly with software and content. "This becomes especially apparent with reference to the rapid developments in smartphone technologies" (Oh and Larson 2011, 171). As discussed in chapter 3, "The introduction of the iPhone into Korea's market came later than in most other countries and created a distinct iPhone shock. Part of this shock was the realization that the success of the iPhone had much less to do with the phone itself and everything to do with the number, variety, and quality of the applications that consumers could use with the device. It was the software application that caught the imagination of consumers and caused them to dramatically increase their use of mobile data services in Korea" (Oh and Larson 2011, 171).

In the era of the smartphone, the process of globalization is uneven in many parts of the world. The recent growth and global penetration of Korean smartphones, at least for a while in the early 2010s, does not mean that the asymmetrical power relationship between the U.S. and non-Western countries has been resolved, because it is still American platform technologies corporations, including Apple and Google, who wield the dominant power. Without advancing software created in other countries, it is impossible to change the contours of American dominance, because the U.S. is still controlling the global markets through operating systems in the realm of the smartphone. "Economic value, power, and ideology are dimensions of all social relationships and should by no means be understood as isolated levels or crystallized stages" (Gonzalez 2000, 108).

With the rapid growth of Korea's ICTs, in particular smartphones, we might argue that the local force becomes a key player in the globalization process; however, since Korean-based smartphone devices have lost their grip in the global market, including the U.S. market, it is prudent to wait to announce the increasing role of the emerging market. U.S. smartphone industries, both smartphone devices and operating systems, continue their dominance in the global market. As new technologies appear, a few non-Western countries are able to develop and compete with Western corporations; however, as far as the software sector goes, the gap between these two areas is deepening. Therefore, interconnectedness via ICTs, including smartphones, cannot guarantee the flattening of the world economy and culture.

5

App Economy in the Digital Platform Era

Applications have changed people's daily activities. In particular, Koreans easily download mobile games and enjoy them on their smartphone apps; when they move around by subway, they simply check their Seoul subway app to find locations; and as food delivery in Korea is among the best, Koreans just click through the apps to deliver food to their homes; most of all, Koreans must have Kakao Talk—the nation's top free instant messenger app to easily connect with family and friends. Applications in Korea have become the lifeblood of its culture and economy.

The phenomenal growth of the smartphone in terms of the number of users and its functions has changed the notion of the digital economy and culture. While the Internet is still significant in the digital economy and culture, the increasing use of both smartphones and applications (apps) ushers in the development of a new digital economy, which is an app economy. In the era of smartphone, smartphone users rely on applications to communicate with friends, play games (both mobile and social games), check news and information, and share ideas with online community members. Likewise, mobile apps are currently being used to check weather, read e-books, listen to music, watch videos, and enjoy many other playful aspects of daily activities. Consequently, they are how "business is done, how employees are connected, how consumers share, learn, and buy. Every business is becoming an applications business. Every industry is becoming an application-centric industry, and the business model shift is only accelerating. We all truly live in an app economy now" (Chambers 2013).

While Western countries are eager to advance app technologies, the growth of apps in Korea is distinctive because of their astonishing use in several fields. The new apps developed by Korean-based ICT corporations as well as those imported from other countries have made people's daily lives fruitful, while contributing to the growth of the app economy. Of course, the increasing role of the app economy has caused new forms of commodification and monopoly capitalism. Smartphone users are spending a disproportionate amount of their time and attention engaging with mobile apps, and therefore, platforms including Facebook and Instagram, as well as Kakao Talk, are better positioned to monopolize time and attention (Marvin 2015). Applications such as WhatsApp and Kakao Talk are indeed gaining hundreds of millions of users, and as businesses, they experience not only faster growth but also speedier monopolization than has ever been seen (Natanson 2015), a phenomenon that asks us to evaluate the significance of the commodification of users and monopoly capitalism.

This chapter explores several vital aspects of the app economy and their implications for the ICT-driven national economy in Korea. As Cheng (2012, 49) points out, it is crucial to investigate "the social system embedded within the new media/app industry driven by increased Internet mobility and public use of smartphones." The chapter therefore examines the rapid growth of apps in the socioeconomic milieu specific to Korea, which is itself becoming part of the global app economy. It also investigates a handful of social-economic issues embedded in the app economy, including the commodification of app users and monopoly capitalism, which will eventually hurt the app economy unless it is properly checked and amended.

Understanding the App Economy in the Smartphone Era

Many ICT corporations around the world have developed diverse applications in order to attract smartphone users. Because of the significance of apps, as Google Play and the App Store exemplify, people access these applications, instead of the mobile Web, to enjoy their daily activities. As Rainie and Wellman (2012, 107) point out, 10 years ago "mobile phones were only for talking, chatting, and snapping, but people rapidly started using their mobile devices to access the Internet." This is when the value and impact of mobile connectivity became most pronounced. While mobile connectivity is a social lubricant, there has been a "boom in apps that have turned smartphones into diversified personal and portable computing devices that can access the Internet," resulting in the remarkable growth of the app economy (Rainie and Wellman 2012, 107).

In fact, until the mid-2000s, the Internet and the greater telecommunications system within which the Internet has been intertwined comprised a leading edge of the epic growth of economic activity (Schiller 1999; 2007). The focus is shifting. Whereas people accessed the Internet through the Web in the 1990s and the early 2000s, many people now access the Internet through applications on the smartphone, and therefore, they are critical products and services in the digital economy and culture in the early 21st century. The digital economy is "moving away from a primarily search-based mentality on desktop to a mobile medium in which typing and search are not the first actions a user takes. Touch is now a user's entry point to app-based, not search-based, discovery" (Marvin 2015). In the app economy, there are several core companies, such as Apple, Google, and Facebook, that maintain a platform on which apps can run.

The proliferation of smartphones has indeed created "an app-centric global marketplace, ushering in the app economy that is driving new business models and revenue streams across all industries" (Marketwire 2012). In the past several years "the mobile industry has experienced a powerful upheaval sparked by the launch of the iPhone and the creation of the first true app ecosystem. This event brought about a gradual restructuring of the mobile value chain and a steady shift in value from the traditional pillars of the mobile economy, telco services and mobile handsets, into app ecosystems." This emerging component of the value chain refers to the "mobile app economy" (Voskoglou 2013). In other words, the app economy, referring to the range of economic activity surrounding mobile applications that encompasses the sale of apps, advertising revenue within apps, and digital goods on which apps are designed to run, has become a large, lucrative business in just a few years (MacMillan and Burrows 2009). The remarkable growth of apps underscores that people are now rapidly going online using smartphones, which demonstrates the presence of a new pattern of digital convergence with smartphones.

The app economy lends itself to being understood by examining several types of metrics. First, it counts the number of apps in a particular app store, how many different developers, and how many times apps have been downloaded (Mandel 2012). This suggests that the app economy is important because it creates new jobs. In this regard, Nick Dyer-Witheford (2014, 127) points out, "A new and enigmatic figure has recently appeared in North America's anxious dreams about jobs, prosperity, and the very fate of global capitalism: that of the app worker. It is estimated that there are more than half a million software application workers in the U.S. alone, many of them developing software for mobile devices."

Second, it is important to find revenues from "app downloads, in-app revenues, sales of virtual goods, and sales of physical goods and services" (Mandel 2012). Third, the app economy is not the same as our traditional information economy, because the app economy primarily relies on the number of users, as well as the volume of sales. In other words, while the previous information economy relied on direct marketing, the current form of the app economy is based on the number of users, which has triggered the growth of advertising as one of the major revenue resources for apps. Apps cannot be flourished under the same economic model (Levie 2013). Therefore, it is critical to understand the diverse economic impacts that apps have contributed to in our economy and culture.

Growth of the Global App Economy

The app economy has become one of the most significant fields in the digital economy. The diffusion and use of smartphones and apps have shaped economic growth, business performance, and subscribers (Jin 2014, 166). "Application services have existed for some time now and should be part of any discussion of media and information technology and industry" (Cheng 2012, 49). However, the app economy is still in its infancy, because Apple launched its App Store in July 2008, at the same time that iPhone 3 went on sale. Development was relatively slow in the first few months, and apps were not something that Apple heavily promoted (Goggin 2011a). Nevertheless, Apple gave the app industry for mobile devices a jump start, particularly for tablet PCs, iPads, and the iPhone and other smartphones (Cheng 2012).

Within two years of being established in 2008, the app industry led to the development of a new type of market economy and became the "process" of the era, with the greatest potential to change the software market since the advent of Microsoft's Windows visual operating system. Just like the dot-com-based World Wide Web, the app system has distinguished itself from the traditional Internet system at both the economic and the visible market level, making it possible for a subsystem to emerge. "As a newly emerged system, the app industry has become a major player in the market based on a few important external indicators." The app industry is capable of "attracting capital and spreading information" (Cheng 2012, 49).

More specifically, the launch of the iPhone created the first true app ecosystem (Voskoglou 2013). This emerging component of the value chain, called the app economy, represents the fastest-growing area in the mobile sector today and will continue to be so in the foreseeable future. As Goggin (2011a, 151) points out,

> There exists a bewildering array of apps available across a number of apps stores and handset types. These apps themselves have wrought a metamorphosis in our notion of mobile phones and media. An app can make it possible to imagine and do things with a mobile phone that were previously never associated with the technology.

Indeed, "The apps developed for App Store follow a specific style and they needed to pass an approval process. There is an economic split between Apple and developers for apps. In this way Apple was cultivating a two-sided market by selling the device and the content. Those who bought an iPhone were also required to buy a specific data subscription" (Ling and Svanes 2011, 9).

Against this backdrop, the global app economy has substantially grown. In 2012, the global app economy accounted for 18% of the combined app services and handset market. By 2016 the app economy rose to about one-third of the combined market, equivalent to half of the handset market. As Voskoglou (2013) reported, "VisionMobile has developed a model for sizing the direct app economy that uses a large-scale, fine-grained data set obtained via Developer Economics surveys. This model incorporates not just revenues generated directly through apps but economic activity generated via commissioned app development, mobile app e-commerce as an app monetization model (i.e., not including e-commerce as core business), venture capital (VC) funding, services for app developers, and several other income sources directly related to mobile apps." According to this model, the global app economy was worth $53 billion in 2012. It grew at a 28% CAGR (compound annual growth rate) between 2012 and 2016, reaching $143 billion in 2016. The bulk of this growth came from Asia and Latin America, the fastest-growing markets in terms of smartphone penetration (Voskoglou 2013) (fig. 5).

Of course, the phenomenal growth of the global app economy has been made possible primarily because the development of apps running on smartphones is a relatively recent phenomenon: the app economy started from almost nothing. When Apple's iTunes App Store and Google's Android Market launched in 2008, smartphone users could choose from about 60,000 apps. On March 20, 2010, the App Store had over 150,000 different applications, and the total number of application downloads had

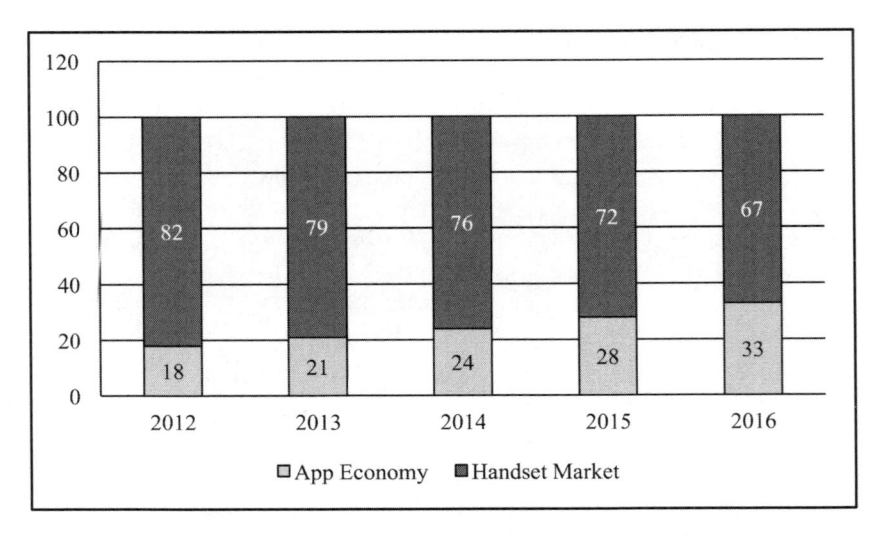

Fig. 5. App Economy Forecast, 2012–16 (Unit: %). (*Source:* Data from DevelopersEconomics.com 2013.)

reached 3 billion (Manovich 2013). As of March 2013, there were more than 827,000 apps in the App Store and about 670,000 apps in the Android Market, and Blackberry had over 40,000 applications, which consumers could access from a variety of mobile devices, including smartphones and tablets (Organization for Economic Cooperation and Development 2013a, 8). Apple's online operation had become entrenched over the course of a few years. Apple's App Store led the pack with a soaring quantity of apps, though it faced stiffening competition from Google and from other app stores in China and Korea (Schiller 2014).

In particular, Apple's App Store (as with its iTunes Store) provided a common distribution mechanism across all operators and Apple devices. This eventually provided "Apple unprecedented economies of scope and a large unified market for its developers" (West and Marc 2010). This rapid app adoption has led to changes in the way that people access information and content on mobile phones and portable devices such as tablets.

One of the reasons that apps have generated so much interest is that they are the first powerful software platform that people carry with them nearly all the time (Organization for Economic Cooperation and Development 2013a). As a reflection of the growth of the app economy, during his keynote speech at the Interrop 2013 conference in New York, Cisco CEO John Chambers talked about how technology trends are moving toward a focus on applications. Chambers said, "We're moving from a web economy to an application economy" (Kerner 2013). Both app developers and advertisers are keen about the potential of apps, perhaps now already proven by apps' enormous contribution to the global economy.

Korea's Emerging App Economy

A few key features characterize the Korean app economy. Korean users gravitate toward several key apps, such as web portals, mobile games, and free instant mobile messengers. From transnational corporations to small start-ups, including Kakao Talk Inc.—which turned into a commercial giant—they have jumped onto the app bandwagon because apps have become lucrative businesses ever since Apple launched the Apple Store based on the huge success of iPhones (MacMillan and Burrows 2009). As the country has vigorously developed its own mobile telecommunications system, Korea has simultaneously developed a new form of economy, based on applications on the smartphone, primarily through having the highest penetration rate in wireless broadband (as well as wired broadband), smartphones, and free mobile messaging systems. In addition, as one of the most active nations in online games, Korea has changed its focus once again toward mobile games, although online gaming has been the largest (see chapter 7). This implies that Korea has created new jobs, increased its export of smartphones and mobile games, and developed locally based search engines, which results in the growth of the app economy. As in many other countries, there are many smartphone services that rely on apps because they give the users the opportunity to keep their smartphones updated with the latest apps and services (Jin 2014). However, many Koreans use apps more intensively than people in other countries, and the changing behavior of Koreans as consumers and/or users consequently intensifies the role of the app economy.

Snowballing of Apps in Daily Cultures

As many apps are available for smartphone users, apps usage in Korea is among the highest in the world. Korean smartphone users had the largest number of regularly used apps (median of 55), followed by Singapore (47), China (16), and Japan (11) (Nielsen Asia 2012). In detail, according to Nielsen's mobile consumer report published in March 2013, Koreans' smartphone use is unique. About 81% of smartphone users used apps at least once a month, much higher than in countries such as Brazil (74%), China (71%), the U.S. (62%), the UK (52%), and India (13%). Korea also ranked first in several categories, such as mobile banking, instant messaging mainly due to Kakao Talk, and video/mobile TV viewing, as table 2 shows. These data help support the idea that Korea is the most advanced in terms of apps usage in the world.

If one looks at app categories, Korean smartphone customers primarily use their gadgets for games and communications. In 2013, among those who downloaded apps, the most popular type was mobile gaming (63.9%), followed by music (43.6%), news (28%), and audiovisual software (including movies). In terms of their use, mobile gaming (58.4%) and communications (42.6%) are the two most common types of apps among Korean users (Korea Internet and Security Agency 2013) (fig. 6). As for communications apps, Kakao Talk, Line, and Skype are popular because users continue to rely on the same mobile messenger app, and these communication apps are some of the most popular types. Accordingly, the majority of smartphone users in Korea enjoy apps for entertainment and other daily activities to get necessary information, such as maps for navigation.

As will be fully discussed in chapter 7, while Korea's major gaming product continues to be online games played on computer servers, the market for games played on mobile devices like smartphones and touchscreen tablets has increased faster than expected (Baek 2014). This indicates that mobile gaming is going to be the fastest growing sector in the near future. It also reflects the rapid uptake of smartphones, and it appears that the trend will continue because smartphones have become one of the most significant ways to access social media for almost everyone.

TABLE 2. Activities Performed by Smartphone Users at Least Once a Month (unit: %)

	SMS	Mobile banking	E-mail	Social network	Apps	Streaming music	Instant messaging	Video / mobile TV
Australia	94	40	55	58	59	21	33	19
Brazil	85	28	66	75	74	39	57	43
China	84	42	58	62	71	59	67	39
India	45	7	17	26	13	11	15	8
Italy	89	22	51	47	49	26	35	17
Russia	95	32	55	59	64	41	34	36
Korea	93	51	52	55	81	40	70	44
Turkey	78	4	33	69	38	22	50	9
UK	92	28	68	63	56	20	37	19
US	86	38	75	63	62	38	28	28

Source: Data from Nielsen 2013, 21–22.

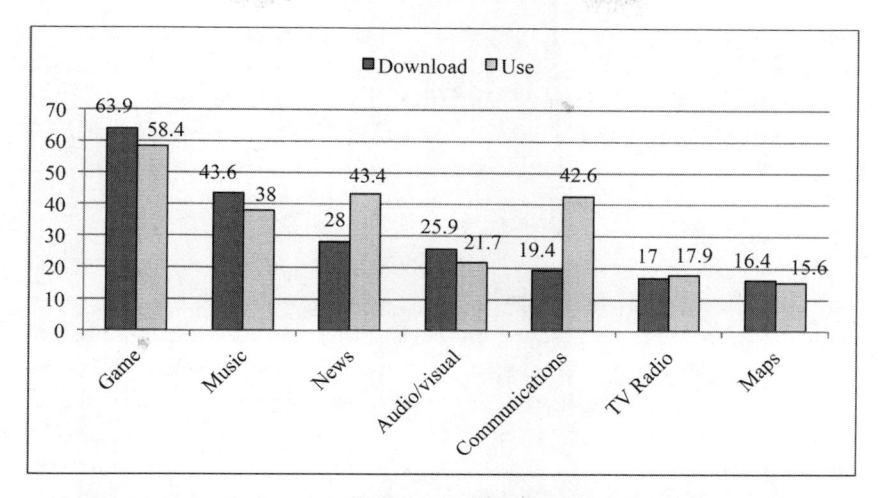

Fig. 6. Most Popular Apps for Korean Smartphone Users in 2013 (unit: %).
(*Source:* Data from Han'guk Intŏnet Chinhŭngwŏn [Korea Internet and
Security Agency] 2013.)

Search Engines on the Smartphone

One of the most significant areas that appeared in tandem with the app
economy is locally based search engines, or web portals, such as Naver,
Daum, and Nate. Web portals are crucial for most Koreans, who use them
to read newspaper articles, play games, listen to music, and enjoy webt-
oons (*wept'un*)—a web comic that is typically published in chapters[1]—and
movies. Web portals have become important parts of the app economy.
Among these, Naver has become the primary player in the portal market,
as its market share in 2014 was 81.6%, followed by Daum (14.9%), Google
(1.9%), Zoom (0.9%), and others (0.8%, including Bing and Yahoo) (Internet Trends 2014). Given the dominant role of a few domestic search engines like Naver and Daum, global search engines, including Google,
Bing, and Yahoo, are not key players in the Korean market.

According to Statista (2014), Google's dominance in the global search
engine market has continued over the last several years. Although Google's
market has been slightly reduced, from 90.6% in July 2010 to 88.2% in July
2014, its power undoubtedly continues. However, Google's position in Korea is marginal. There are several reasons for this weakness. First, it does
not have enough Korean-language data to satisfy Korean customers. Second, it does not properly cater to Korean culture, which is "a collectivistic
culture that places a strong emphasis on relationships" (Asian Correspon-

dent 2010). This means an individual's preferences are often determined by public opinion, and Koreans perceive "the Web as a place where people gather to discuss and share information." As a result, "to be successful in Korea a web service needs to reflect what topics are currently on people's mind to avoid potential embarrassment for users. But the interface of Google focuses on individual personalization that presumes each individual has different preferences and styles. So Google has put all the services relevant to personal preferences behind the interface to present the common function Search" (Asian Correspondent 2010). Koreans use Internet portals, including Naver and Daum, to enjoy several functions, including mobile games, GPS, information, and weather; however, global leaders, including Google and Yahoo, have focused on information search, which has rendered them invisible to users. With the exception of operating systems that Android and iOS dominate, smartphones and other apps are primarily produced by Korean IT corporations, whether megagiants or start-ups.

However, when we go to the mobile search engine, the story is somewhat different because Google has continued to maintain its strong standing. According to KoreanClick (cited in Y. Kim 2014), unlike the Internet search engine market, Google is extending its dominance in the mobile sector, although in many cases Naver has been the major search engine in the mobile sector in Korea. In January 2014, for example, 19.3 million people visited Google through mobile, while 16 million did Naver and 12.8 million on Daum (fig. 7).

Unlike the Internet area, in the mobile arena these three search engines are relatively competitive with each other in Korea. The chief reason for Google's lead is that Google's Android plays a key role. Google's Android is the largest operating system, and it fully utilizes its advantage. When people open the smartphone, Google is located on top of the screen, which makes it easy to use, while Naver and Daum need users to download applications (Y. Kim 2014). While the forgoing data may be somewhat of an outlier, because other data collected early in 2015 show that Naver is the largest player in the mobile search market, the numbers show the significant role of Google in the mobile search market, compared to the Internet search market (Y. L. Choi 2015).

Meanwhile, as a reflection of the importance of the usage of apps, mobile web use has declined. According to Flurry—a mobile marketing firm (Perez 2014)—users in the U.S. spend 2 hours, 42 minutes per day on mobile devices as of March 2014, up from 2 hours, 38 minutes as of 2013. Mobile app usage accounts for 2 hours, 19 minutes of that time, while mobile web usage

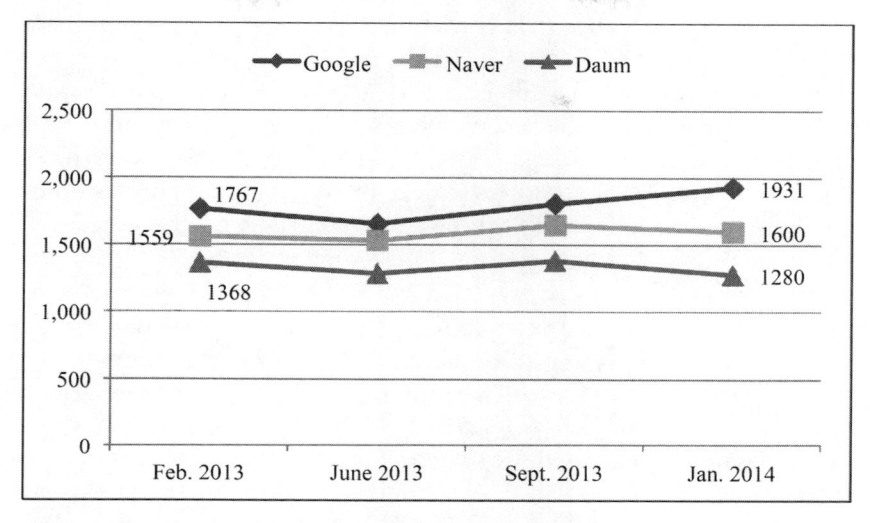

Fig. 7. Mobile Search Engine in Korea (Unit: 10,000). (*Source:* Data from Y. Kim 2014.)

dropped from 20% of the U.S. consumer's time in 2013 to just 22 minutes per day as of March 2014. In December 2011, people around the world used their smartphones evenly between mobile app (50.7%) and mobile web (49.3%) (Nielsen 2012b). Although many Koreans enjoy smartphones for web searching, more than 90% of Koreans in 2012 used them for apps, including mobile messaging, games, and maps. This implies that the Korean mobile ecosystem relies on apps, instead of the Web, contrary to the global trend (Jin 2014).[2] By comparison, U.S. smartphone users spent 89% of their mobile media time in the fourth quarter of 2013 using mobile apps, while only 11% of their smartphone time was spent using the mobile Web (Nielsen 2014). While the majority of countries are still using the mobile Web as their main gateway to access the Internet on the smartphone, several leading countries in the realm of smartphones and relevant apps have rapidly increased their use of apps in their daily activities.

The App Economy in the Workplace

On an economic level, many apps represent jobs for programmers, for user interface designers, for marketers, for managers, and for support staff. "The app economy was responsible for roughly 466,000 jobs in the U.S. in February 2012. This total includes jobs at 'pure' app firms such as Zynga, a San Francisco–based maker of Facebook game apps that went public in December 2011. App economy employment also includes app-related jobs at large companies such as Electronic Arts, Amazon, and AT&T, as well as app 'infrastructure' jobs at core firms such as Google, Apple, and Facebook. In addition, the app economy total includes employment spillovers to the rest of the economy" (TechNet 2012). The app economy in the workplace is soaring. For example, Apple (2015) announced that it created more than 1 million jobs as of January 2015. By breakdown, applications designed for Apple's smartphones and tablets have helped create more than 627,000 U.S. jobs and 66,000 people work directly for the company in the U.S. Of course, as the *New York Times* (Wingfield 2012) argued a few years ago, Apple slightly exaggerated the numbers because it included the drivers who deliver devices as well as the people who build the trucks that get them there; however, it is good enough to prove the importance of the app economy in creating new jobs.

Compared to this, in Korea in 2013, about 276,600 worked in the mobile Internet industry, a 4.6% increase from the previous year. Among these, the system/gadget field accounted for 55.2%, followed by content development (18.7%), networks (16.2%), and solution/platform (9.9%). However, only 2,586 employees worked in mobile application-related areas among the content development area (Korea Mobile Internet Business Association 2014, 346). This implies that Korea has increased the number of jobs related to the app economy, which is noticeable; although the country needs to develop software and relevant workforces. As the software sector is not comparable to a few advanced economies, Korea cannot truly benefit from the growing app economy.

However, the growth of the app economy does not automatically guarantee the advancement of the national economy. The growth of the app economy causes the collapse of some pre-app era industries, because apps replace these technologies and services. For instance, only a few years ago, GPS navigation systems were cutting-edge technologies; however, with the emergence of apps like SK planet T Map and KT Olleh Navi, as well as free road apps, the GPS navigation producers and operators have been forced to shut down their operations. Several GPS navigation corporations have experienced a 20%–30% decrease in revenues per year in their market, and the number of companies has greatly decreased (An and Ku 2014).

The newspaper industry has been hit as well: first, because of the Internet, and second, because of smartphones and apps. As people use mobile

Naver and Daum to read newspaper articles (previously on the Internet), the newspaper industry has lost advertising revenues and readership. In January 2015, one online newspaper reporter indeed officially announced on Facebook that he terminated his subscription to a newspaper that he had subscribed for several years because he now reads articles online; he does not need to have newspapers that he does not read anymore. According to the Korea Press Foundation (2014), newspaper corporations, including sport newspapers and free newspapers, experienced a dramatic decrease in their revenues (up to 50%), primarily due to smartphones.

The significant role of mobile messenger applications has fundamentally shifted power from telecommunications services corporations to instant mobile messenger application operators. Texting apps are costing wireless carriers, which for years relied on these services for the bulk of their revenue. Texting apps had taken away $23 billion in revenue from carriers as of the end of 2012, according to market research firm Ovum (Rusli 2013). Just as the music industry almost collapsed after the emergence of the Internet, some industries cannot deal with the app economy. This means that the app economy, in terms of the creation of new jobs, is based on the sacrifice of other industries, and that is why we need to carefully determine the future of the national economy. Although it is significant to develop new jobs in app-related areas, we also need to deal with the new job situation in nonapp areas, which is a new socioeconomic issue in our contemporary app economy.

Instant Mobile Messenger Applications as a Corporate Sphere

The app economy has several distinctive characteristics. Most of all, the current app economy differentiates itself from the general notion of the information economy because its primary benefit lies not in the intrinsic value of apps (Melody 2009). "Unlike previous ICTs, smartphones function as intermediate goods and services, which result in hundreds of thousands of apps. The growth of the app economy will be further intensified when application software and multimedia content can be provided easily and reliably through smartphones" (Jin 2014, 175). In particular, instant mobile messenger applications, such as WhatsApp (U.S.), Viber (Cyprus), WeChat (China), Line (Japan), and Kakao Talk (Korea), are taking primary roles in the app economy. These mobile messaging apps are some of the most significant applications in the contemporary app economy, because they act not only as a voice messaging system but also as a platform for users to enjoy value-added services, including mobile games.

Through instant messenger applications developers and corporations have advanced their unique business model based on an increasing number of subscribers. Several significant monetization strategies are available for them, including selling apps, providing advertising, and creating them as profitable commodities. These mobile messenger applications corporations have turned themselves into symbols of "a corporate sphere," which amplifies their market value (van Dijk 2012), and they are the most desirable digital commodities in the app economy.

There have been several important transactions in this area in recent years. In 2014, Rakuten, a Japanese Internet firm, paid $900 million for Viber, founded by Israelis but based in Cyprus. Alibaba, a Chinese online giant, paid $215 million for a slice of Tango, another Silicon Valley firm, in March 2014. Tencent, Alibaba's rival, owns WeChat, which has almost 400 million users. It also runs QQ, an older messaging service, and has a stake in Kakao (*Economist* 2014). Several search portals, telecommunications corporations, and social networking sites, including "Facebook, Deutsche Telekom, and Samsung, are also making a land grab into the messaging app market. Facebook acquired mobile messaging start-up Beluga in 2011" (Rusli 2013). In February 2014, "Deutsche Telekom invested $7.5 million in messaging app Pinger," while SK Telecom, Korea's largest mobile carrier, purchased MadSmart, maker of a popular messaging app in 2013. "Samsung, which debuted its own messaging app called ChatOn in late 2011, has reached out to mobile messaging startups including MessageMe for possible partnerships" (Rusli 2013).

Meanwhile, Facebook acquired WhatsApp in 2014, which is second in terms of transaction value only to HP's acquisition of market leader Compaq (Hartung 2014). Messaging apps, known as "over-the-top" apps because consumers do not pay carriers directly for the services, "typically go beyond plain vanilla text messaging with features such as voice notes, camera applications, and access to games and virtual stickers that are popular among teens and young adults." In January 2013, "WhatsApp said it processed 18 billion messages a day, up from 10 billion five months earlier. The app had more than 100 million downloads on Google's Android phones alone" (Rusli 2013). Facebook finally finished its acquisition of WhatsApp for $22 billion in October 2014 (Frier 2014).

Of course, mergers and acquisitions in social media as part of digital platforms have been common because of their significance as profitable commodities based on user bases. In fact, Google acquired YouTube for

\$1.65 billion in 2006 (Google 2006), and News Corporation bought MySpace. Microsoft also acquired LinkedIn for \$26.2 billion in June 2016 (Mims 2016). While it is not unusual to witness significant mergers in the media and telecommunication sectors, these recent transactions are very interesting because the deals happen between new media and new media, in particular, social media institutions. Social media corporations heavily relying on users' daily activities appropriate their cutting-edge applications in order to attract more users, which consequently make them valuable commodities in the deal market as well as the stock market.

In Korea, Kakao Talk was a main target because of its increasing number of users, which makes the mobile instant messenger app profitable. Kakao Corp., operator of Korea's dominant mobile messaging service Kakao Talk, merged with smaller listed Internet portal Daum Communication in an all-stock deal that valued the messaging app operator at \$3.4 billion (Song 2014).[3] The merger between Daum and Kakao Talk occurred right after Facebook's acquisition of WhatsApp. The merger gave Kakao's shareholders the lion's share of the new company, although Daum had more revenue, profits, and staff, plus a stock market listing. The deal shows that messaging apps are hot property (*Economist* 2014). Daum doubtlessly hopes that Kakao Talk, which was installed and registered on 145 million devices at the time, helps it combat Naver, Korea's leading portal. "Koreans do not just use the app to chat: it is also a popular platform for mobile games," from which Kakao makes most of its money, and "for sending both digital and physical gifts. Naver, too, owns a messaging app, Line, with 400 million users, but it is based in Japan" (*Economist* 2014).

Instant mobile messaging services, again, have been hot commodities in the app economy. Because of their substantial role, several ICT corporations, from telecommunications to search portals, and again to social network sites, have attempted to have mobile messaging service corporations. As the Internet has been a target for many transnational corporations, instant mobile messaging service companies are the most wanted corporations in the 21st century app economy. In the realm of instant mobile messenger applications, the industry has been corporately converged into a few transnational corporations.

The App Economy as a Symbol of Monopoly Capitalism

While the increasing role of the app economy in Korea is not necessarily negative, one major issue in the current app economy is the concentration of network power in a few corporations. It is not unusual for market power to concentrate into a handful of mega firms, and the smartphone production business and the wireless telecommunications market also show an oligopoly market. There are three major monopolistic oligopolies in Korea's app economy: the instant mobile messenger market, the operating system market, and the apps stores market. While the first market has been controlled by Korean mobile messenger applications, the two other markets have been dominated by American-based apps, which result in increasing inequalities between the U.S. and Korea, as in many media and telecommunications sectors.

To begin with, in the Korean mobile messenger market, as of April 2015, Kakao Talk (87.4%, in terms of the market share) was the leader, followed by Line (3.2%), WeChat (1.8%), and MyPeople (1.6%%) (Y. Chang 2015). Kakao Talk's leading position slightly decreased from 92% in 2014 because of surveillance issues; however, its position did not change (*Associated Press* 2014).[4] In the operating system landscape, Android, which was invented in 2003 and integrated by Google, has been the world's best-selling smartphone operating system. Android accounted for 85.4% in terms of the market share, followed by iOS (14.1%) and others (0.5%), during the first half of 2014. Likewise, in the portal market, the market share of Naver exceeded 81.5% between January and August 2014 (Y. Pak 2014).

Thus, the dominant network industry characteristic in the app economy is a highly concentrated network monopoly market. This raises a concern for governments "with respect to the application of existing competition laws and/or direct industry regulation" (Melody 2009, 69). As different digital technologies emerge, the market has rapidly moved from an oligopoly market to the monopoly market, although it is not in a perfect monopoly state, which is a new trend in the app economy. In the oligopoly state, a few key players compete with each other, and their market share is relatively even. However, in the app economy, one superpower controls the entire market because of its absolute market share, sometimes hovering around 90%, while other competitors together account for less than 10%. What is significant in the app economy is that progress toward the monopoly market is not the same as in old forms of capitalism. As McChesney (2013, 37) correctly observes,

> Pure monopoly, in which one firm sells 100 percent of a product and can scare away or crush any prospective competitions, almost never exists. Instead, capitalism tends to evolve into what is called monopolistic competition, or oligopoly. These are markets where a

handful of firms dominate output or sales in the industry and have such market power that they can set the price at which their product sells.

In many industries, including auto, retail, and oil, "Under oligopoly there is strong disincentive to engage in price warfare to expand one's market share, because all the main players are large enough to survive a price war" (McChesney 2013, 37).

However, in the app economy, the monopolistic dominance is not about a price war, but the war over the number of users. "The more users, the better profit"—that is the business motto for the app economy. Registering on Kakao Talk is free, and downloading apps is also mostly free. Search apps themselves are free, as is the case of the operating system on the smartphone. Therefore, the common sense of the monopoly system and/or the oligopoly market cannot be applicable in the app economy. Instead, these superpowers earn profits as people increase their use of apps, because they benefit from advertising revenues. This new trend asks us to critically comprehend the nature of the app economy, which has aggravated existing social inequality. The apps are recognized as supposedly wonderful new technologies that encourage equality in human communication and information sharing; however, the app economy based on smartphones and apps service has already experienced several negative consequences (Jin 2014). As Daubs and Manzerolle (2016, 65) argue, "The emergence of what we call app-centric media—a move away from an open Web and toward small, stand-alone applications exacerbated by the rapid uptake of smartphones and mobile broadband use—continues the commercialization of mobile connectivity."

American Powerhouses in the Korean App Economy

In the Korean app economy, the increasing role of American app-related technologies is noticeable, resulting in American dominance in the local market. According to the Korea Mobile Internet Business Association (2014), the Korean mobile content market was expected to be worth $3.18 billion in 2014, up from $2.43 billion in 2013. The mobile content market includes fee-based apps and advertising in apps. In 2013, sales of apps as a form of fee-based download was the largest among business models at 40.8%, followed by in-app purchase options wherein a user can complete transactions within the app (20.4%), sales + in-app purchase (11.3%), sales + advertising (8.6%), advertising (5.9%), and others (0.4%). The sale of apps is the largest category because e-book and multimedia, as some of the most significant app contents, are largely purchased through direct marketing (Korea Mobile Internet Business Association 2014, 17–18).[5]

Google Play alone took 49.1% of the mobile content market, followed by the App Store (30.5%), which means that these two Western-based platforms accounted for 79.6% of the Korean app economy. For example, in the mobile game sector in 2013, the majority of Koreans directly downloaded mobile games from Google Play (67.1%), followed by Kakao Talk (49.9%), Telecommunications Service Providers (20.9%), and Apple App Store (18.6%) (Ministry of Culture, Sports and Tourism 2014, 442). This trend will continue, primarily because Android holds more than 90% of the domestic operating system market. In other words, smartphone users easily access Google Play when they buy their new smartphones (Nam 2014). Since Google Play or the App Store is preinstalled on smartphones, they have natural advantages, and they continue to increase their market share.

In fact, it is very difficult for users to install apps that are not preinstalled, because Google prevents other app stores, including Naver and the App Store, from registering on Android. Article 4.5, the noncompetition clause in Google's Developer Distribution Agreement, clearly explains that "users may not use the market to distribute or make available any product whose primary purpose is to facilitate the distribution of software applications and games for use on Android devices outside of the market" (Google 2014). Users need to take at least 12 steps to download and use uninstalled apps, which technically blocks their use. This implies that Korean developers and smartphone service providers and developers have serious difficulties in extending their market share in the domestic market.

During the third quarter of 2014, Korea ranked fifth in the number of downloads on Google Play, only behind the U.S., Brazil, India, and Russia. However, in terms of revenue for Google Play, Korea ranked third. Games have accounted for the majority of Google Play's worldwide revenue. Nearly all the revenue growth for Google Play in Korea came from gaming (App Annie 2014; Crawley 2014). As Gerald Goggin (2011a, 154) points out, Apple also "seeks to bind consumers to its handset (iPhone, iPad, and iPod touch); which to do basic things such as purchase and upload software or digital media (music) must be used in conjunction with its digital management and rights system (iTunes); which in turn only offers software—apps—approved by Apple, or otherwise these cannot be distributed via the apps store. The iPhone and iTunes have been the subject of much criticism regarding their enactment of restrictive regimes of intellectual property and user control."

Accordingly, only a handful of U.S.-based platform corporations have benefited from the growth of the app market, which is not uncommon in our platform-driven market economy. Korea is going to be one of the major app economies, as with previous key ICT sectors, including broadband services, the Internet, online gaming, and smartphones; however, software-driven apps are not comparable to these previous new technologies and services, because the smartphones have to rely on American-based operating systems (Jin 2015). Android and iOS are the symbols of U.S. platform technologies, software in particular, and Korea cannot make its own counterparts to these operating systems. Unless Korea develops its own operating system, Korea's app economy will be limited and/or marginalized.

Unlike other areas in the app economy, including smartphones, free mobile messengers, search engines, and mobile games, the operating system on the smartphone has yielded a unique ecology, because two U.S.-made operating systems, Android and iOS, control the domestic operating system market. Although Samsung and LG are key players in smartphone markets, they have no operating systems, and they have no choice but to use foreign-based operating systems.[6] Korea depends primarily on Android. Since Samsung and LG use the Android OS as their operating system, an increasing role for Android in the Korean smartphone market is inevitable. Both the iPhone apps and Google apps platforms are "premised for the most part of the dominant interests of commercial industry—with app stores opening up a new market in the interstices of mobile networks still heavily controlled by dominant transnational mobile carriers in alliance with handset vendors, and new intermediaries, creating tightly coordinated value chains" (Goggin 2011, 153). The power of apps creates new resources and new productive capacity, which is dramatic, and its strategic uses have great benefits for app developers and app-owning corporations (Couvering 2012); therefore, current monopolistic capitalism (at least the duopoly of Android and iOS) continues to extend inequalities between the U.S. and other countries.

Several foreign corporations also directly invest in Korea's app-based venture capital. For example, in November 2014 Goldman Sachs invested $36 million in Woowa Brothers (Uahan Hyŏngjedŭl), a Korean start-up that operates the country's most popular food-delivery mobile service. Woowa Brothers, whose "Baedal Minjok" (Paedal ŭi Minjok), or "Delivery Nation," service processed about 4 million food-delivery orders from 145,000 registered restaurants in October 2014, recorded about 470 million Korean won ($424,000) in net income on $9.7 million in revenue. Baedal Minjok conducts almost all of its business through mobile devices—about 99% of its transactions take place on a smartphone in Korea (J. Cheng 2014). As smartphones and apps have changed people's daily activities, including food delivery, foreign financial and ICT corporations have rapidly invested in Korea, increasing the dominance of Western-based transnational corporations in the Korean app economy.

The economic structure of societies, moving away from manufacturing and toward service industries, has been changing since the early 1970s (White 2014). Most of all, the digital economy based on the growth of the network society (Castells 2010) has been accelerated since the late 1990s with the advancement of the Internet. The early 2010s witnessed the rapid shift from network society connected through the Internet to network society connected through both smartphones and apps. The emerging app economy has rapidly replaced the old information economy because apps on smartphones are fundamental resources for the growth of the digital economy and culture.

Conclusion

This chapter has discussed the app economy, which has shown unique growth in tandem with smartphones. The app economy has become an area that policymakers and telecommunications practitioners evaluate because of the increasing role of apps in the digital economy and culture. Since the introduction of the smartphone, over a million apps have been developed and hundreds of billions' worth of apps downloaded. There is an app for almost everything. Many governments and start-ups envisage an app opportunity in terms of jobs and GDP. They regard the emerging app economy as the next driver of innovative growth. Many countries share an ambition to become the most innovative economy in the world, and the app trend is regarded as an embodiment of opportunity.

This outcome, of course, is no surprise. The uptake in smartphones has increased the significance of apps, as they bring more values into the smartphone market (M. Jung 2010). Apps have substantially contributed to the growth of our digital economy, in creating jobs, enhancing revenues for app developers and corporations, and raising the GDP (gross domestic product) (Warmerdam 2014). Dramatic improvements in smartphones are changing the ways in which "knowledge is generated and communicated, and thereby the ways that firms operate, markets function, and economies develop" (Melody 2009, 93). Smartphones act not only as a new electronic communication foundation but also as infrastructure for the app economy.

Applications are crucial in the early 21st century, not only because they change people's daily activities in stock transactions and mobile gaming, but because they substantially contribute to the growth of the contemporary economy embedded in the digital economy. Apps are directly generating the social and cultural range of the capitalist economy as never before; therefore, it is appropriate to call this form of capitalism the app-driven digital economy.

As with the rapid acceptance of the smartphone, Korea has emerged as one of the powerhouses in the app economy. In addition to two global leaders in smartphones—Samsung and LG—Korea has its own search portals, including Naver and Daum. Korea's instant mobile messenger applications, such as Kakao Talk and Line, play a key role as platforms that applications can run on. Korea has also swiftly developed mobile games as video game users have switched their platforms from online to mobile games. One missing point in Korea's lineup of capacities is its own operating system on the smartphone. Both Samsung and LG use Android as their sole operating system; therefore, Google Play has become a major platform in the app market. This unbalanced system causes concerns, in particular, with respect to the increasing inequalities between the U.S. as an operating system holder and Korea as an operating system user.

In sum, the rapid growth of the app economy implies the continuing dominance of U.S.-based platform corporations, including Google and Apple, primarily due to their leading role in operating systems on smartphones. As Korea has increased its reliance on American operating systems, both Android and iOS, in addition to several social networking sites, including Facebook and Twitter, the app economy cannot avoid Korea's relative marginal status. While Korea has rapidly developed its hardware, including smartphone gadgets, and several mobile games, Korean firms have no choice but to use Google Play or Apple App Store. In the era of platform technologies, the lack of software and relevant apps is a critical problem for the Korean app economy. Therefore, developing software, in particular operating systems, is the key to growth in the app economy.

6

From Digital Divide to Digital Inclusion in the Smartphone Era

Mi-yŏng Kim (age 21), a college student, upgraded her smartphone from a Samsung Galaxy 3 to a Galaxy 5 in December 2014. When she switched phones, she also upgraded her service plan to LTE Data Unlimited, which means that her parents pay more than 100,000 Korean won per month for value-added applications. At the same time, Mi-yŏng's mother (age 54) still uses her feature phone. Her service plan is the cheapest, and she pays around 35,000 Korean won per month. With her feature phone, she primarily talks with family members and friends; however, she cannot watch K-pop music contests and movies, unlike her daughter, because she can access the Internet only through Wi-Fi, not through the data plan. (notes from interview with a twenty-year-old female student in Seoul)

Since the mid-1990s, Korea has rapidly developed several ICT areas, such as broadband services, video games, wireless telecommunications, and smartphones, as well as related applications. As Koreans quickly adopt these technologies, these ICTs greatly influence the national economy and youth culture. It has been widely believed that technological advancement would "improve living conditions in an almost utopian way." However, technological advances have brought with them new challenges, "to which every society has had to adapt" (Garcia de la Garza 2013). Korea, as one of the most networked societies, has experienced a number of setbacks in the smartphone era due to sociocultural issues, including an ongoing digital divide, cyberbullying, and privacy invasion.

Among these, the digital divide has become a serious social issue in the smartphone era. Although many policymakers and ICT experts believe that smartphones bridge the digital divide, the reality is that the smartphone has not become the solution. While the digital divide related to the Internet has been reduced, many people still do not benefit from smartphones, primarily because of soaring subscription fees or a lack of skills. Many housewives and the elderly, as well as some high school students, cannot subscribe to new and expensive smartphone services, nor are they able to access several value-added applications. Inequalities in people's access to and use of smartphones and social media, such as Facebook and Twitter, cause economic inequality, not only as technologies but also as symbols of the knowledge economy.

This chapter analyzes the digital divide pertinent to smartphone technologies. It contextualizes the digital divide in ways that go beyond statistical measurements. It examines the ways in which the digital divide can be understood in the smartphone era; it discusses major issues in the smartphone divide, and suggests a shift of emphasis, from a traditional understanding of the divide to a dual divide, which is a new sociocultural issue in the discourse on the digital divide. In order to achieve these goals, in addition to a historical approach contextualizing the recent growth of the smartphone and relevant sociocultural issues, I use survey research on "people's consciousness of smartphones," which was conducted in Korea in December 2014. A total of 1,000 mobile users were interviewed; respondents were 19 years of age or older, and of the sample, 49.5% were male and 50.5% female. Among interviewees, 81.8% had a smartphone, while 18.2% did not, instead using feature phones. This hybridized methodological framework stands to contribute much to moving research and inquiry forward in smartphone studies, and in particular, in analyzing current sociocultural issues behind the rapid growth of smartphones in Korean society.

How to Understand the Digital Divide: The Problematic of the Binary Divide

Since the mid-1990s, when the notion of a digital divide became a major policy issue in countries around the world, addressing the digital divide has become one of the most significant agendas in many countries. IT policymakers and scholars started talking about the digital divide when personal computers made their way into households in the 1990s, a time that included a growing Internet and the World Wide Web. "While a cadre of people in the U.S. trumpeted the significance of the Internet and moved forward on National Information Infrastructure deliberations and pronouncements, scholars and critics grew increasingly alarmed at disparities in using computers across certain populations" (Strover 2014, 114–15).

As the Internet has become one of the most important information and communication technologies, the digital divide generally refers to discrepancies between social groups based on level of income, education, employment, race, ethnicity, and age in access to, use of, and empowerment by networked computers and the Internet (van Dijk 2005, 2006, 2012a, 2012b) and/or mobile phones (Chircu and Mahajan 2009; Lee and Kim 2014; Y. J. Park 2015; Mascheroni and Olafsson 2015; Puspitasari and Ishii 2015). In other words, the notion of the digital divide was initially popularized in connection to the disparity in Internet access (Castells 2001; Gunkel 2003). Policymakers have had difficulties in resolving barriers such as access, skill, and infrastructure, which make the conditions of social minorities even worse (Quan-Hasse 2013). Two major problems were previously identified in the discourse of the digital divide: one was inequalities in material access to technologies, and the other was inequalities in the skills to use ICT efficiently (Selwyn 2004).

Several previous works documented the acquisition of computers and their use in home, work, and school, and investigated the demographic elements that predicted ownership and use. In so doing they helped to define the digital divide as a matter of physical access to technology. Based on their empirical research, they claimed that women, minorities, and the elderly, and more generally the poor and the less educated, were indeed using computers less often. With these studies, "The prescriptive or aspirational role of computers and Internet use, the idea that one would need these skills and technologies in order to function within the information society, entered the mainstream" (Strover 2014, 115). The digital divide measures the gap between those who are empowered to substantially participate in an information and knowledge-based society, and those who are not (Kaplan 2005, cited in Mancinelli 2007). Some of them have argued that the Internet reinforces social inequality already established in the social structural division (Norris 2001; van Dijk 2006).

Of course, those narrowly defined notions implying a bipolar division between the haves and the have-nots, as well as the connected and the disconnected, cannot explain the multiplicity of recent dynamics (Mansell 2002; Warschauer 2003; 2004; Selwyn 2004; Livingstone and Helsper 2007; Kim et al. 2011). These scholars argue that "mere access is insufficient to ensure equality of opportunity, seeking to move the debate from a concern with material access to the technology to the trickier question of social and cultural factors that influence use" (Livingstone and Helsper

2007, 672). As Selwyn (2004, 349) points out, "a lack of meaningful use of them [ICTs] is not necessarily due to technological factors (such as a lack of physical access, skills or operational abilities), or even psychological factors (such as a 'reticence' or anxiety about using technology), as is generally claimed by technologists. Instead, as a range of studies have shown, individuals' engagement with ICTs is based around a complex mixture of social, psychological, economic and, above all, pragmatic reasons." Meanwhile, Tsatsou (2011, 323) argues that "the linear, simplistic and normative character largely overlooks the role of sociocultural and political capital and the importance of their connections for how people adopt ICTs and for social implications of ICT adoption."

The idea of the binary divide is indeed problematic, because it is clear that the underlying framework of the digital divide is technological determinism—the view that the mere presence or absence of a technology has a determining effect on behavior and social development (Warschauer 2003, 297). "Technological determinists range from those who see media's impact as automatically good to automatically bad, but they agree on the overriding role of technology in determining social change. This results in an emphasis on bridging the technological gap through the distribution of hardware, software, and online networks" (Warschauer 2003, 297–98). In this regard, Warschauer points out that

> [the] big problem with the digital divide framing is that it tends to connote digital solutions, i.e., computers and telecommunications, without engaging the important set of complementary resources and complex interventions to support social inclusion, of which informational technology applications may be enabling elements, but are certainly insufficient when simply added to the status quo mix of resources and relationships. (298)

With the recent growth of smartphones, the digital divide in the early 21st century has remained worrisome. As smartphones have taken a primary role in the networked society, the digital divide cannot be the same as that of the 1990s when the concept was originated with the emergence of the Internet, because of smartphones' huge potential, not only for economic advantages, but also for sociocultural opportunities (Jin 2015). As Castells (2001, 3) indeed pointed out, "The digital divide was the concept based on the Internet-based economy and culture"; however, the smartphone era has created another significant form of the digital divide. Therefore, it is necessary to understand the digital divide as an effect of the various causes and consequences that our society witnesses with the growth of smartphones.

As Mascheroni and Olafsson (2015) point out in their recent article, research examining mobile telephony, in particular smartphones, in terms of "digital divides has been sparse compared to the body of writing on digital and social inequalities associated with Internet access and use." Since the early 2000s when the mobile phone became an important social communication tool and a multifunctional medium, however, inequality in possession and use of mobile media, which is the mobile divide, has drawn greater attention (Lee and Kim 2014). In particular, Korea, because of its phenomenal smartphone use, requires that we critically deliberate on the digital divide in order to reflect on the milieu surrounding the growth of smartphones, not feature phones. Simply focusing on access or skill ignores a smartphone's capabilities to perpetuate inequalities because smartphones are not used in the same way as the Internet. When scholars and government policymakers started to think about the role of the Internet in tandem with the digital divide, one of their major interests was whether people could benefit from information they acquired from the Internet; however, smartphones have more diverse functions with many more applications than the Internet. The discussions on the digital divide should be rather nuanced, reflecting diverse standards for measuring disparities in tandem with the growth of technologies, including smartphones, than previous works (Jin 2015). Newly developed conceptual frames are needed to deal with the contemporary digital divide occurring primarily due to discrepancies in the prevalence of smartphones.

From the Internet Divide to the Smartphone Divide

The digital divide in many countries is perhaps less stark today than it was in the late 1990s (Benkler 2006). As growth rates of access among underrepresented groups are higher than the growth rate among highly represented groups, the disparity between the haves and the have-nots in technology access, which constitutes the basic digital divide, has been somewhat reduced. As the access gap shrinks, it becomes less relevant in discussions of social inequalities (Tranter and Willis 2002, cited in Ji and Skoric 2013). After analyzing the digital divide in the U.S., Compaine (2001, 334) indeed argues that "it is fair to propose that the digital divide is disappearing on its own. Public policy in a few years can then turn its attention to the much smaller skirmishes that may be needed to help out with the digital crevice left at the fringe."

Several indicators also point to a lessening digital divide based on the access to and use of ICTs. One of the recent evaluations comes from the International Telecommunication Union. According to the ICT Development Index (IDI), which is a composite index combining 11 indicators into one benchmark measure that serves to monitor and compare developments in ICT across countries, Korea is a top ICT country. The IDI was developed by ITU in 2008. The IDI is divided into three subindices, including an access subindex, use subindex, and skill subindex (International Telecommunication Union 2013).[1] The results of the 2012 ICT Development Index show that there are major differences in ICT levels between countries. In 2012, IDI values ranged from a low of 0.99 (Niger) to a high of 8.57 (Korea) (table 3). As one of the goals of this index is to determine the digital divide around the world, this certainly indicates that the digital divide in Korea is lower than in other countries. Through the rapid growth of information and communication technologies, Korea has

TABLE 3. ICT Development Index (2012)

Rank	Country	ICT Development Index
1	Korea	8.57
2	Sweden	8.45
3	Iceland	8.36
4	Denmark	8.35
5	Finland	8.24
6	Norway	8.13
7	Netherlands	8
8	United Kingdom	7.98
9	Luxembourg	7.93
10	Hong Kong	7.92
11	Australia	7.9
12	Japan	7.82
13	Switzerland	7.78
14	Macao	7.65
15	Singapore	7.65
16	New Zealand	7.64
17	United States	7.53
18	France	7.53
19	Germany	7.46
20	Canada	7.38

Source: Data from International Telecommunication Union 2013, 24.

reduced the digital divide. However, this index only uses people's access to and use of ICT; therefore, it explains only the binary digital divide. Of course, we cannot deny that the digital divide with regard to basic access and skills is still important throughout the world.

More significantly, the digital divide in the realm of smartphones is just getting started, which is a new concern in the networked society, because smartphones are not only about technological disparities but also about sociocultural inequalities. In this regard, Sparks (2013, 39) argues, "If economic and social inequalities are among the key determinants of the digital divide in all of its manifestations, these have certainly not been significantly reduced, and in some important cases have increased, as the Internet and social media are undergoing development and diffusion."

The use of smartphones is critical in the early 21st century because the smartphone is not only a mobile device but also a handheld computer (Verkasalo et al. 2010). People can search for information, download documents, and use diverse services on a smartphone. Smartphone users are able to install many applications and utilize them to fulfill their own interests. SNSs are also loaded on a smartphone as one of the major applications (Hwang and Park 2013). The relative affordability of smartphones has made them the bridge across the Internet's long-discussed digital divide, "opening up the possibility that some of the inequalities of access that existed prior to the advent of smartphones might be bridged" (King 2011; E. Park 2014, 1). Taking an example of American young adults between 18 and 19 years of age, Hargittai and Kim (2012, 26) indeed argue that

> mobile devices are often heralded with the potential to help underprivileged groups leapfrog limitations they may have in accessing the Internet through other means. Our findings suggest caution in this realm given that it is precisely those people who have more Internet experiences who are also more likely to use mobile devices for more functionalities including online access. Consequently, it remains to be seen whether the spread of cell phones can bridge the digital divide or whether the adoption of devices with different functionalities and affordances will exacerbate the existing gaps between information haves and have-nots.

Such a smartphone gap leads to a dual digital divide (Kim et al. 2011; Selwyn 2004): an intergroup divide between smartphone users and nonusers, and an intragroup divide among "smartphone users caused by differences in skill levels, permitting some to enjoy more sophisticated and advanced usage" (E. Park 2014, 1). The differential among smartphone users arises because the smartphone is a versatile multimedia platform. While smartphones and feature phones share some functionalities, smartphones have more capabilities, such as a camera, touch screen, GPS navigation, Wi-Fi, and mobile broadband access.

Above all, the open environment of smartphone applications enables the creation of functions that were never intended. The more skilled a user is in operating the smartphone, the greater is the possibility that he or she is able to "fully exploit the technical capabilities of the device" (E. Park 2014, 1). Therefore, an intragroup smartphone divide arises between users who are confined to a limited set of functions, and users who are able to use a diverse set of applications. Accordingly, in a converged and smart-media environment, it no longer makes sense to talk only of a digital divide based on access to a platform—instead, a new "smartphone divide" is created based on a user's ability to access and use an array of different services (E. Park 2014). Many people cannot benefit from the smartphone era.

The smartphone is a new breed of ICT that has not yet received much attention from researchers because of its short history. Although several previous works paid attention to mobile technologies, their main focus was feature phones, not the current form of smartphones. We therefore can evaluate some of the major characteristics of the digital divide and apply them to mobile phones in order to determine the nature of the digital divide in the smartphone era. For example, as Chiru and Mahajan (2009, 458) point out, the smartphone divide can be better understood by focusing on two distinct divide factors: (1) smartphone technology depth (which measures the level of penetration, or adoption), and (2) smartphone technology service breadth (which measures the variety of smartphone services available for adoption). Smartphone depth is defined as the penetration of smartphones, irrespective of provider (measured as the number of smartphone subscriber accounts). The digital divide should also be measured along an additional metric, smartphone technology service breadth. Smartphone communication technology is not single-purpose but can be used for a variety of tasks such as voice and data transmission (mobile text messaging, e.g. Kakao Talk, email and Web browsing, downloading applications, and streaming media). Smartphone service breadth is defined as the service variety, or number of different services available to smartphone users (Chircu and Mahajan 2009, 458). Thus, it is possible through these measures to capture the reality of smartphone use in Korea, where smartphones serve not only as communication devices but also as information, entertainment, and Internet access channels.

The Smartphone Divide in Smartland Korea: Toward a Dual Digital Divide

Korea has witnessed an ongoing digital divide in the smartphone era. The disparity of smartphones is rather conspicuous compared to the PC-based digital divide. Whereas the penetration rate of smartphones has been increasing, smartphones yield a new form of digital divide because many people cannot afford them, resulting in the marginalization of non-smartphone users in terms of their ability to access information and applications. More importantly, many people do not have value-added services because soaring monthly subscription fees during the economic recession period in the early 2010s priced them out of the market. According to *OECD Communications Outlook 2013*, Korea's monthly household expenditure on communication equipment and services was $148.30 on average as of 2011, ranking third behind Japan ($160.50) and the U.S. ($153.10) among 26 nations where related data were available. The expenditure figure includes fees for the Internet access and mobile and fixed-line communication. However, of the expenditures of Korean households, that for mobile communications was $115.50, which was the highest among the surveyed countries, followed by Japan ($100.10), Mexico ($77.40), and Finland ($77.10) (Organization for Economic Cooperation and Development 2013b, 278) (fig. 8). This means that some Koreans cannot afford to subscribe to this expensive service.

In Korea, the digital divide in terms of access to and use of ICTs has been relatively reduced. The use of ICTs among social minorities, including disabled, low-income, agriculture/fishing, and the elderly, in 2013 was 75.2%, up from 45% in 2004, and from 71.1% in 2010, assuming the level of information of the general public is 100% (Ministry of Science, ICT and Future Planning 2013). This means that the level of information of social minorities has substantially improved, although about 25% of them did not benefit from ICT infrastructure. However, the digital divide among mobile users and nonusers (the mobile divide) is much less than the PC-based divide. The mobile divide was 42.9% in 2013, although it has been elevated from 27.8% in 2012 with the rapid growth of the smartphone era (table 4). Mobile access is the percentage of mobile owners who are able to access the Internet, and mobile skills are the combination of skills and competence; therefore, it is the level of digital literacy.

Digital use here includes both quantitative use and qualitative use. Quantitative use is whether mobile users actually use the Internet through mobile devices and considers the amount the time used, while qualitative

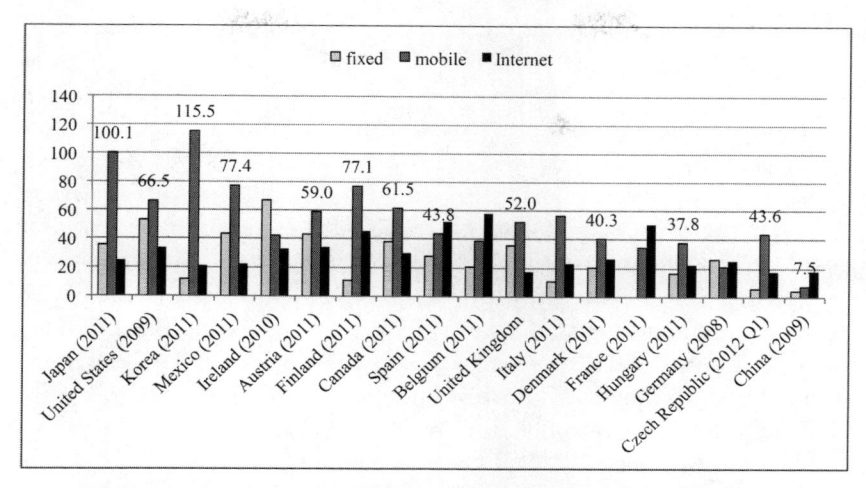

Fig. 8. Monthly Household Expenditures on Communications in OECD (unit: US dollars). (*Source:* Data from OECD 2013, 278.)

use is whether the mobile use helps people's daily activities (Ministry of Science, ICT and Future Planning 2013, 366–69). Mobile access is relatively better than mobile skills and mobile use in that the number of smartphones continues to grow. About 57.5% of social minorities have access to the Internet through mobile phones; however, mobile skills were recorded at 36.2%, while mobile use was 36.8%. These data imply that social minorities need to develop their digital literacy and mobile use in order to benefit from the wide spread of mobile technology, including smartphones (table 5).

Korea seems to have achieved the highest rate in terms of smartphone use; however, the dual divide has emerged, as Kim et al. (2011) point out. The first component of the dual divide is that between those who have smartphones and those who do not have smartphones. Since about 73% of

TABLE 4. Digital Divide in Korea, 2013 (Unit: %)

	Disabled	Low-Income earners	Farmers / Fishermen	The Elderly (above 50s)	Total
PC-based divide	83.8	83.2	67.8	72.6	75.2
Mobile-based divide	41.8	63.8	35.7	38.8	42.9
Smartphone penetration	39.9	55.1	35.7	41.5	42.8

Source: Data from Ministry of Science, ICT and Future Planning 2013, 366–67.
Note: Average is weighed based on the number of people in each category.

mobile users were using smartphones instead of feature phones as of September 2015, this simple linear divide will disappear in a few years, just as Korea experienced with broadband services.

As proven with several previous new media developments, including broadband, online gaming, and the Internet, one of Koreans' primary characteristics is demanding quick change, which expedited the growth in these cutting-edge technologies. Once regarded as calm and patient, Koreans demanded quick change, a mentality that became distinctively Korean. Since the 1960s when the country began its high-speed transition from an impoverished state to one of Asia's major economies, Koreans have been impatient. Koreans are not known for taking things slowly. Instead, Korean society is known for its impulsiveness—desiring quick communication, quick contact, and not waiting (Jin 2011). There is no exception in smartphone use. As their peers start to subscribe to the smartphone, Koreans must have it.

Under this circumstance, the divide among smartphone users has become a new concern. The divide is not only caused by differences in skill levels, permitting some to enjoy more sophisticated and advanced usage, but also caused by the disparity in terms of either subscribing to expensive plans for value-added services or subscribing to cheap plans for basic services. Although it is important to understand disparities in skills among smartphone users, it is also critical to understand disparities in people's ability to buy different services.

The dual digital divide in the smartphone sector is more systematic than that which measures differential access to the Internet. Ostensibly, users subscribe to the cutting-edge service; however, many of them do not enjoy significant services that others have access to. Seen in the wider sense, the concept of digital divide is related to the relative position that an individual or a group has in the whole context of the society. A divide may be due to education, ethnicity, or location. In any case, there is a link be-

TABLE 5. Mobile-Based Divide in Korea, 2013 (Unit: %)

	Disabled	Low Income	Farmers / Fishermen	The Elderly	Total
Mobile access	53.5	74	48	55.8	57.5
Mobile skills	37.2	58.9	31.1	30.7	36.2
Mobile use	35.8	59.9	29.2	32.1	36.8
Total	41.8	63.8	35.7	38.8	42.9

Source: Data from Ministry of Science, ICT and Future Planning 2013, 368–69.
Note: Average is weighed based on the number of people in each category.

tween poverty and a digital divide, but poverty is not the only factor in this regard (Mancinelli 2007).

The Smartphone Divide as a New Social Issue

According to my survey, the smartphone divide in terms of access, which is the basic and first form of the dual divide, clearly exists in several categories. First, by age, virtually all people who are working and/or studying in college and are under 49 years old have at least one smartphone. However, people who are in the next decade of life (50–59) or older do not generally have smartphones. Only 33.8% of those over 60 have a smartphone. When they retire, their income dramatically decreases, and they have to save by reducing unnecessary spending, including the smartphone service subscription. Second, the survey result shows that while university (95.6%) and postsecondary graduates (92.9%) have the smartphone, elementary (18%) and middle school degree holders' (35.8%) smartphone subscription rate is much lower. Finally, office workers (98.4%), retail/service workers (95.2%), and professionals (90.5%) are relatively higher in their smartphone subscription rates than housewives (66.7%), simple laborers (47.1%), agriculture/fishing/forestry workers (45.5%), and those retired or are unemployed (43.2%) (table 6). This result proves that the smartphone divide is a critical issue even in terms of access, because social minorities (e.g., the elderly, blue-collar workers, and less-educated people) are using the smartphone less often. As will be detailed later, this systematic digital divide among smartphone users combines with the divide stemming from the quality of use, which is the second component of the dual divide.

The second-tier divide is typically outlined according to a few variables: skills, abilities like online navigation and problem-solving abilities, and empowerment—that is, for example the effectiveness of use, social relationships, and the extension or growth of social capital (Molnar 2003). DiMaggio et al. (2003; see also Molnar 2003) also suggest that more educated people use the Internet, which empowers them to acquire further social benefits, including social capital. DiMaggio et al. (2003; see also Molnar 2003) draw a very important conclusion: as a result of different Internet use patterns, some people (e.g., the more affluent) will multiply their skills (empowerment), which further increases social differences.

The digital divide in the smartphone era creates more complicated and multifaceted consequences than the Internet divide, because smartphones are the core gadget in people's daily activities. Therefore, it is critical to understand the smartphone divide in two different layers: one is the divide between the smartphone users and the no-users, and the other is the divide within the users—value-added service users and basic-service users. In particular, the new type of digital divide, which is the second form of divide within users, is significant because it yields several sociocultural discrepancies, including differences in civic participation and cyberbullying, and the gap between a few corporations and general users.[2]

TABLE 6. Do You Have a Smartphone? (Unit: %)

Categories	Samples	Yes	No	Total
Total	1,000	81.8	18.2	100
Gender				
Male	495	83.4	16.6	100
Female	505	80.2	19.8	100
Age				
19–29	181	98.9	1.1	100
30–39	191	100	0	100
40–49	212	96.2	3.8	100
50–59	197	86.3	13.7	100
Over 60	219	33.8	66.2	100
Highest level of education (certificate)				
Elementary	50	18	82	100
Middle school	95	35.8	64.2	100
High school	474	86.7	13.3	100
University	367	95.6	4.4	100
Graduate school	14	92.9	7.1	100
Job				
Management	9	77.8	22.2	100
Professional	42	90.5	9.5	100
Office worker	184	98.4	1.6	100
Retail/service	168	95.2	4.8	100
Mechanic	43	74.4	25.6	100
Simple labor	34	47.1	52.9	100
Agriculture/fishing/ forestry	22	45.5	54.5	100
Business owner	189	82	18	100
Housewife	204	66.7	33.3	100
Student	67	98.5	1.5	100
Retired / no job	37	43.2	56.8	100
Other	1	100	0	100

Sociocultural Interpretations of the Digital Divide in the Smartphone Era

Korea is among the leaders in the OECD in the total ratio of household expenditure that people devote to communication devices and services. This reflects the rapid adoption and take-up of smartphones and mobile broadband. Korean consumers use their smartphones heavily, with Cisco data suggesting an average use of 1.2GB per month in 2011. This monthly usage is significantly higher than in other countries (OECD 2013, 278). Korea's expenditure on the mobile sector will be greater in the future as smartphones continue to replace feature phones.

On the flip side, many Koreans are not able to subscribe to smartphone services; if they have a smartphone, they subscribe only to basic services. The second tier of the dual divide can be found in several areas in Korea. To begin with, while 30% of mobile subscribers still used feature phones as of September 2014, about 10% (4 million) of smartphone subscribers used mobile virtual network operators (MVNOs) that started their services in mid-2011. The MVNO borrows networks from mobile carriers at cheap prices, and in turn lures consumers with mobile services that are lower priced than those offered by mobile carriers. MVNOs, including e-mart, FreeT (Pʼŭri Tʼellekʼom), and Korea's Postal Office, offer mobile rates that run 30% to 50% below those of major carriers. MVNOs offer services with limited calling time, number of messages, and data. While this kind of service helps to grow smartphone penetration rates, MVNO subscribers are not able to enjoy several value-added services.

More specifically, while they can access the Internet through Wi-Fi, which does not count against their plans' data limits, they have only limited access to the Internet and/or apps in Wi-Fi-less zones, because their plans offer insufficient data limits. In addition, many people still use 2G/3G as their gateway to the Internet, although LTE has rapidly grown. When the customers limit their service because of soaring wireless bills, they cannot enjoy the same level of value-added service, which results in the digital divide. Data service is primarily needed for value-added services, including mobile games and media streaming, and MVNO subscribers and/or those subscribing to lower-priced plans cannot enjoy these audiovisual entertainment and information features, which is a new form of digital divide in contemporary Korean society. Since the number of MVNO subscribers has rapidly increased, from 2.12 million in September 2013 to 4.0 million in September 2014, and again to 5.6 million in September 2015 (Ministry of Science, ICT and Future Planning 2015a), the divide among smartphone users has been intensifying. The most significant issue in this intragroup divide among smartphone users is the restriction on data usage that limits MVNO subscribers. In turn, disparities among value-added service subscribers and basic-service subscribers among smartphone users are increasing rather than diminishing.

Second, the survey shows that the subscription fee is the most important element when people choose their smartphone service plan, resulting in the smartphone divide. About 50.4% of people surveyed responded that the service plan (subscription fees) is the single most significant factor, followed by convenience (24.9%), data capacity (12.5%), design (11.2%), and others (1%) (fig. 9). By age, younger people worried less about subscription fees than older people. While 38% of 19–29 year olds worried most about subscriptions fees in selecting their smartphone service, 55.9% of 50–59 year olds, and 71.6% of people over 60, worried about subscription fees the most. In particular, only 9.4% of participants between 50 and 59 and 4.1% over 60 years of age stated that they cared about data capacity; it is the subscription fee that is crucial for almost all of the people over 50. Consequently, many older people, blue-collar workers, and some students cannot have the value-added services, which require payment of soaring subscription fees. Indeed, among interviewees, 62.3% responded that they were overwhelmed by current subscription fees, while only 12.6% did not feel pressured. Others (25.1%) said they were indifferent. There are no significant differences in this category by age, location, and job, which means almost all of the participants feel similar pressure from the current smartphone service plans.

Another indicator also echoes this reality. When they were asked "whether fee-based applications were burdensome or not," 47.4% of participants said they felt burdened, while 23% of them said they did not feel burdened. About three-tenths (29.6%) said that they were indifferent. Interestingly enough, the age group 19–29 felt the most pressured (52.5%), while 44.6% of those above 60 felt some burden. College students and people in the early stages of their careers are heavy users of apps, and they have to download some fee-based apps in order to enjoy their cultural activities. However, since they are either students or job seekers in many cases, they felt financial difficulties in consuming apps.

These data provide evidence that the elderly do not subscribe to expensive service plans; therefore, they mostly use their smartphones for basic functions, instead of benefiting from value-added services. The younger customers are not much different in terms of pressures related to subscription fees. In general, people feel financial pressure subscription fees rise as

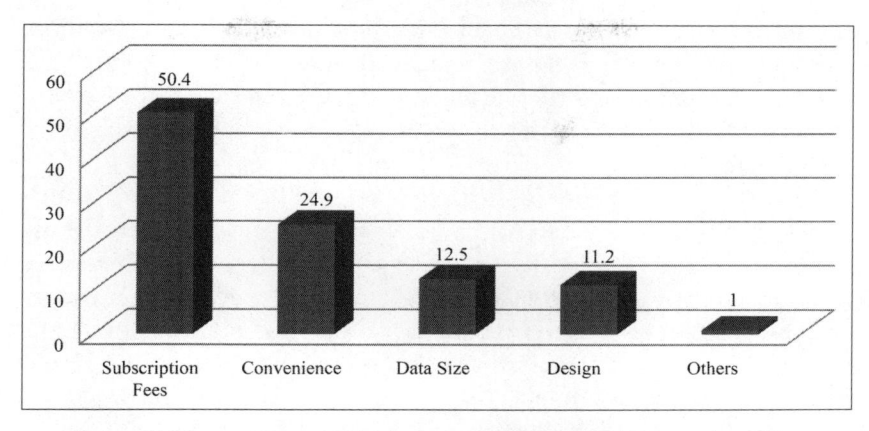

Fig. 9. What Is the Most Significant Factor in Selecting Your Smartphone Service Plan? (unit: %)

smartphone makers advance new technologies and add new value-added services. What is significant is that this kind of pressure may cause limited smartphone usage, which restricts people's economic activities. In this regard, Warschauer (2004, 7) has already pointed out that

> the notion of a digital divide implies a chain of causality: that lack of access (however defined) to computers and [the smartphone] harms life chances. While this point is undoubtedly true, the reverse is equally true: those who are already marginalized will have fewer opportunities to access and use computers and [the smartphone]. In fact, technology and society are intertwined and co-constitutive, and this complex interrelationship makes any assumption of causality problematic.

Third, the skill divide is also problematic, because the majority of Koreans make limited use of the smartphone no matter how expensive the service plan they subscribe to. When we asked about "their ability to use the smartphone," only 22% of respondents said that they fully understand and use every function. About four out of five interviewees make limited use of the smartphone whether they fully understand its functions or only understand basic functions. In fact, 38.8% of them said they make limited use of their smartphone although they know its functions very well, and 33.6% responded that they know only basic functions and use them. A small number of respondents (5.4%) said they only understood and used the mobile-level functions of their smartphone—in other words, those that a feature phone would have. The divide between those who fully utilize the smartphone and those who limitedly use the smartphone is substantial. Therefore, we are able to confirm a substantial second tier of the dual divide.

This is an especially interesting result because this survey implies that many people do not need to have a smartphone because their level of use is restricted. Their ownership of the smartphone does not change their activities much because they use the smartphone as if it were a feature phone. As far as the smartphone divide is concerned, access might not be a serious issue, but their use of the smartphone for better cultural and economic activities is limited.

Consequently, many Koreans believe that the smartphone does not help reduce the digital divide. In response to the question, "Do you think that social media can reduce the digital divide?" 53.2% of participants stated that social media could reduce the digital divide; however, 35.4% said that there were no differences, while 11.4% believed that social media increased rather than decreased the digital divide (fig. 10). By age, 60.8% of younger age group members (19–29) stated that social media could reduce the digital divide, but only 40.6% of older group members (above 60) agreed. By job category, while 64.3% of professionals believed social media could reduce the digital divide, only 32.6% of blue-collar workers and 45.1% of housewives believed in the possibility. This result shows that Koreans attitudes toward the digital divide and its resolution vary depending on their age and job, which demonstrates disparities between the haves (younger, skilled, and professional job holders) and the have-nots (older, less-skilled, and blue-collar workers) in the smartphone, social media era.

Considering the role of smartphones in the digital divide, some say ubiquitous use of smartphones means everyone can access the Internet, but there are reasons to question what smartphones mean for the digital divide. Smartphones are widely popular throughout the country, among rich and poor, educated and less educated alike. However, as the survey data prove, the digital divide in the smartphone era remains, and the situation is not likely to be resolved anytime soon.

In addition, while higher-income households maintain smartphone subscriptions in addition to some sort of fixed broadband access, statistics demonstrate that lower-income households increasingly are using the mobile phone for their Internet access, probably in lieu of a home-based broadband service. This presents some problems. Cell phone-based Inter-

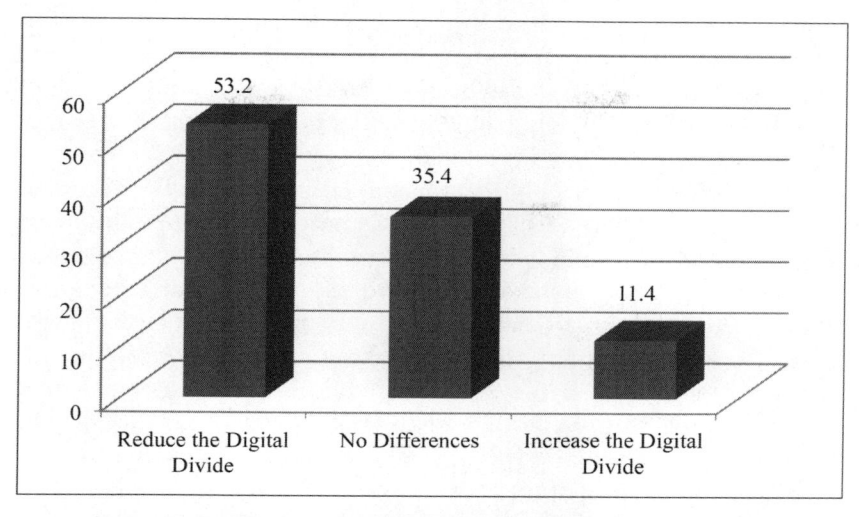

Fig. 10. Whether Social Media Reduce the Digital Divide (unit: % of respondents agreeing)

net access has obvious limitations for tasks such as applying for jobs or even searching for complicated information; the screen size impedes full functionality. As the next generation of wireless devices with larger screens becomes available, and as the speed on wireless networks improves, this barrier may become moot (Strover 2014). However, the cost factors with smartphone-based Internet access are still formidable for low-income individuals, as are usability issues. "As more bandwidth-intensive applications evolve, and if services continue to charge by speed tiers, we are once again facing the prospect of another type of digital divide, one in which higher income households afford high speed mobile services and lower income households are left with less capable services; rural areas likely will be the last in line for the best wireless service" (Strover 2014, 119).

Digital Inclusion in the Smartphone Era: Policy Perspectives

Public policy meant to reduce the digital divide, and in particular the smartphone divide, needs to focus on resolving imminent issues, including the enhancement of both mobile skills and mobile use among these social minorities. The concern in the smartphone era is that exclusion from these smartphone networks is one of the most damaging forms of exclusion in our economy and in our culture (Castells 1997; 2001). As the binary digital divide model had become outdated, policymakers and researchers consider providing an alternate framework, with the goal of digital inclusion, which is a prominent concept in the early 21st century (Warschauer 2002; 2004; Livingstone and Helsper 2007). Digital inclusion refers to the extent that "individuals, families, and communities are able to fully participate in society and control their own destinies, taking into account a variety of factors related to economic resources, employment, health, education, housing, recreation, culture, and civic engagement" (Warschauer 2002). As Quan-Haase (2013, 135) points out, "the most noticeable barriers that exist are not only a lack of 'infrastructure,' 'economic barriers,' 'illiteracy,' 'poor computing skills,' and 'lack of supports,' but also 'cultural barriers.'" Digital inclusion is a matter not only of an adequate sharing of resources, but also of "participation in the determination of both individual and collective life chances" (Stewart 2000, 9). It overlaps with the concept of socioeconomic equality, but is not equivalent to it. "Digital inclusion emphasizes policy intervention to reduce digital inequalities and to foster participation of all citizens to the information society" (Verdegem 2011, 31). In other words, "digital inclusion is part of broader policy measures that aim to include all citizens in society and to foster social cohesion" (Cammaerts et al. 2003, cited in Verdegem 2011, 31).

Since basic ICT competence is a prerequisite for participation in productive online activity, a greater understanding of the contexts of inequality, beyond single factors such as access or diffusion, is imperative (Kvasny and Keil 2006, cited in Mervyn 2014). Therefore, one of the most significant considerations in the smartphone era is to understand the ways in which to enhance the quality of smartphone use. As the majority of smartphone users enjoy basic services, the focus should be advanced use. In other words, while helping the remaining nonusers onto the first rung of the Internet ladder will remain an important challenge to guide policy, for individuals to fully realize the benefits of the smartphone, we must help them move up the ladder—to move from basic activities such as e-mail and browsing to more advanced uses such as e-learning and transactional activities like buying, banking, and accessing government services (Office of e-Envoy 2004, 11, cited in Livingstone and Helsper 2007). The smartphone divide should be understood beyond the technical issues, because without resolving the disparity embedded in sociocultural issues, it is difficult to overcome the digital exclusion related to the smartphone divide.

The increasing use of smartphones has been presumed by some observers to resolve digital exclusion; however, this new development has intensified inequalities in participation. After an empirical study conducted in 2014, Lee et al. (2015, 54) concluded that while the gap resulting from socioeconomic status declines, the gaps in Internet usage and communication competence are greatly increasing. Furthermore, online activities are mainly influenced by the levels of Internet usage. Therefore, they argue, "disengagement from the use of new communication technologies, especially smartphone use, could lead to social, cultural, and economic exclusion" in Korea.

Conclusion

This chapter has analyzed the digital divide in the smartphone era in Korea. It primarily discussed the shift of the digital divide, from a PC-based divide to smartphone-based divide, so that readers may comprehend the significance of the emerging smartphone divide. It addressed the complexity and nuances of several variances in participation that have been newly identified in the smartphone divide. It especially emphasized the emergence of the dual divide in the smartphone era: first, the divide between the smartphone haves and the smartphone have-nots, and second, the divide among the smartphone haves. The chapter argued for the necessity of policy initiatives in resolving the smartphone divide.

The digital divide in the smartphone era has become a new form of digital exclusion and is further marginalizing already oppressed and disenfranchised individuals. This is still an ongoing question. Korea has shown increasing income inequality in the early 21st century. Lower-income households (below 50% of median income) have substantially grown, from 7.1% in 1990 to 10.4% in 2010, and again to 14.5% in 2013 (OECD 2014; *Yonhap News* 2014c). Therefore, although smartphones have been the buzz in recent years, the digital divide will continue.

It might be expected that the smartphone divide will be reduced with the fast diffusion of smartphones in the coming years, if the smartphone divide is about access. However, the digital divide in the smartphone sector cannot be resolved anytime soon, and it will be widening in some sense because of the new form of the divide, a dual divide.

Analogues to the Korean case can be found in other countries. For example, after conducting their empirical research in Indonesia, Puspitasari and Ishii (2015, 10) conclude,

Smartphones have a negative effect on digital divides. A parallel gap between the information rich and poor still exists for smartphones as well as PC-based Internet access. Smartphones are adopted and utilized by younger people along with PC-based Internet use. Digital divides persist even among smartphone adopters. Without higher educational level and younger age, it will be more difficult to adopt and use mobile Internet, even if a person owns a smartphone. Digital divides are differently associated with smartphone and feature phone. Feature phone ownership narrows the digital gap regarding age, income, educational years, and PC-based Internet experience, while smartphones ownership widens the digital gap regarding age, educational years, income, and PC-based Internet experience.

Therefore, it is necessary to consider multifaceted approaches, including the need for a richer, more considerate, and systematic analysis of the ways in which socioeconomic and cultural factors and regulations are interrelated, in order to resolve the smartphone divide (Jin 2015).

It is also vital to develop policy initiatives to reduce the divide for social minorities. In the smartphone era, while focusing on the resolution of the digital divide, we need to extend our perspectives toward digital inclusion because it is vital for people to benefit from their possession of the smartphone. Since they are informed citizens with new media available to them, including the smartphone and social network sites (e.g., Facebook and Twitter), it is imperative for people to use their smartphones to participate in major policy decision processes and social events.

PART III

Smartphones and Youth Culture

7

Transformative Mobile Game Culture on the Smartphone

Mobile gaming has suddenly become one of the most exciting national cultures and lucrative game businesses in Korea. As use of smartphones has burgeoned since 2009, mobile games have caught hold of public consciousness. Until a few years ago, mobile gaming was relatively small and much less popular than online gaming, in particular, compared to MMOR-PGs (massively multiplayer online role-playing games). The story has changed because Korea took a big step toward becoming a "mobile game wonderland" in the era of the smartphone. In other words, the swift growth of smartphones since 2009 has opened up new opportunities for Korean mobile game developers and allowed more people to play mobile games, from casual and board games to role-playing games. As Larissa Hjorth (2011a, 357) points out, "Just as the smartphone encompasses a range of affordances, practices, and modes of co-presence, so too has mobile gaming grown with more possibilities." Furthermore, mobile gaming itself has shifted on a large scale, mainly because the major genre of mobile gaming has been transformed from casual games, such as Anipang (Anip'ang) and Dragonfly, in its early growth in the early 2010s, to mid-core mobile role-playing games, including IDEA and Heroes of Incredible Tales (HIT), released in 2015.

Recognizing the increasing role of mobile gaming in people's daily activities in the smartphone era, this chapter explores the ways in which mobile gaming has become the national pastime in Korea. This chapter examines Korea's distinctive sociocultural factors contributing to the growth of video game culture, in particular mobile gaming on smartphones. The major goal is not to hold up Korea as a nation that determines the mobile gaming future for other countries, nor to suggest that Korea is irreducibly culturally other in its approach to new media technology. Since the sociocultural differences and contextual specificities of mobile gaming cultures are not the same in all countries (Hjorth 2007a; Hjorth and Chan 2009), it is not possible to provide a generalized case for global mobile gaming culture. Rather, by locating Korean mobile games in historical and sociocultural contexts, this chapter investigates the localized nature of global popular culture, which might give some ideas to mobile game users, game developers, and policymakers. It also maps out the transformative mobile game culture in two different perspectives. On the one hand, it discusses the major shift of mobile game genres, from casual games to role-playing games, in order to determine how the newly developed RPGs have changed people's perception of mobile games. On the other hand, it explores whether the recent boom of mobile gaming has been a signal in transforming people's game habits, from online gaming to mobile gaming, or whether Korea will maintain its specific online game culture in the era of the smartphone, which will shed lights on our current debates on the increasing role of mobile games in the most networked society, Korea.

The Evolution of Smartphones as Mobile Platforms

Since the late 1990s, Korea has become one of the most significant markets in the realm of online gaming, with several well-developed MMORPGs, such as Lineage, Lineage II, AION, Tera, and Blade & Soul. Korea's online games have been remarkable both nationally and globally. Domestically, online gaming has been the largest game sector in terms of market share and the number of gamers, surpassing console games, mobile games, PC games, and arcade games early in the 21st century. Globally, Korea's online games have been the largest segment of the export of cultural products. In 2013, Korea exported US$2.71 billion worth of games. The online game industry, as the most significant area, constituted 90% of game exports in 2013 (Ministry of Culture, Sports and Tourism; 2013). Because of the phenomenal growth of Korea's online gaming, both nationally and globally, online game culture is also distinctive. Online gaming has produced mass game culture, the boom of PC-bang (*p'isibang*, Internet café), and daily leisure activities of youth. As Larissa Hjorth (2006) correctly observes, Korea is "an exemplary model of the ubiquity of online gaming culture; highlighting that games and the attendant social spaces and cultural knowledges can be part of everyday lifestyles, rather than a mere leisure activity of a subcultural group."

Unlike online games, Korea's mobile games have been popular since the early 2010s only after the introduction of the smartphone. There were several mobile game corporations, such as Gamevil, Gomid, and Com2uS, that developed mobile games early in the century; however, their role was relatively limited because mobile games were too early for the touch phone revolution, and some of them ended up closing up shop. "Overshadowed by the dominance of online games, mobile gaming had a low profile in the digital game sector." The growth of the mobile game industry and culture was markedly slow until a few years ago. It appeared that the mobile gaming scene had not matched the pace of online PC gaming activities, which were very popular among Korean youth (Jin et al. 2015, 417).

However, Korea witnessed a swift growth of mobile games in the second decade of the century, as the mobile game industry has rapidly developed its own strengths in both domestic and global markets. As discussed in previous chapters, the Korean launch of the iPhone in November 2009 and its quick penetration nationwide heralded the start of Korea's smartphone era and the subsequent development of mobile gaming. "The rapid uptake of smartphones has reshaped the ways in which software developers, users, and academics consider the interrelationship between mobility, culture, technology hardware, and the Internet. In addition, this uptake has added a significant new layer of encrustations around what we might define as 'standard' uses of mobile technologies" (Christensen and Prax 2012, 731; see also Goggin 2009).

Gamers now enjoy diverse mobile games, from casual games to role-playing games. In particular, game users have started to enjoy several role-playing games on the smartphone screen, including Tower of Ascension and Legion of Heroes, which were released in 2013 and 2014, respectively. Enhanced 3D techniques, bigger screens, and visual effects developed in very recent years have become major factors in the growth of mobile role-playing games.

Consequently, the mobile game market has soared. In 2014, the market value of gaming in Korea, including console/handheld, online, mobile, arcade, and PC games, was as much as 87.4 billion Korean won. The online game industry accounted for 63.3% of this figure, followed by mobile (33.3%), console, arcade, and then PC games (Ministry of Culture, Sports and Travel 2013; 2015).

The growth of the mobile game sector is recognizable if one compares it to other types of gaming. Until 2010, console/handheld gaming was the second largest type; in 2011 mobile gaming surpassed console/handheld gaming to become the second largest (table 7). As a reflection of the growth of mobile games, in 2014 when Korea exported $2.97 billion worth of games, the online game sector accounted for 62.4%, while the mobile game industry greatly increased its presence at 36.9% (Ministry of Culture, Sports and Tourism 2015). One of the fastest-growing leisure activities of the 2010s, mobile gaming has quickly developed from a marginalized female players' pastime into a multi-billion-dollar business and one of the most significant forms of popular culture.

While Korea's major gaming product continues to be online games, the market for games played on mobile devices like smartphones and touchscreen tablets has further increased in recent years. The domestic mobile gaming market constituted 11.5% of the global mobile gaming market, second only to Japan in 2013 (Ministry of Culture, Sports and Tourism 2014, 89). This indicates that mobile gaming is going to be the fastest-growing sector in gaming. It also reflects the rapid uptake of smartphones, and it suggests that the trend will continue because smartphones have become one of the most significant ways to access social media for the majority of citizens.

In *Global Mobile Media* Gerald Goggin (2011b, 114) argues that "in a little over a decade, mobile gaming has been established as a significant part of contemporary gaming cultures. It has not proven as lucrative as hoped, but in various forms it is firmly established as a central part of cell phone culture—whether as embedded games on a handset, downloadable games from a portal or premium rate number, or apps." This argument clearly specifies the emerging role of mobile gaming in culture, although it is not yet commercially successful. The mobile sector, however, has suddenly become one of the most lucrative video game markets in Korea, and several major online game corporations, such as NCsoft, WeMade, and Nexon, as well as mobile game corporations, have vigorously developed their new strategies in advancing their mobile games.

TABLE 7. Korean Video Game Market, 2010–14 (Unit: Million Korean Won)

Game Form/Year	2009	2010	2011	2012	2013	2014
Online game	37,087	47,673	62,369	67,839	54,523	55,425
Mobile game	2,608	3,167	4,236	8,009	23,277	29,136
Console game	5,257	4,268	2,684	1,606	936	1,598
PC game	150	120	96	680	380	337
Arcade game	618	715	736	791	825	528
Arcade *pang*	744	768	763	665	639	405
Total	46,464	56,711	70,884	79,590	80,580	87,429

Source: Data from Ministry of Culture, Sports and Tourism 2013, 24; and 2015, 23.

Against this backdrop, two major considerations regarding Korea's mobile gaming culture need to be addressed. On the one hand, the sudden growth of mobile gaming asks us to redefine the notion of mobile games. As is well known, mobile games are played on any portable device, including a mobile phone, smartphone, PDA, or tablet (Richardson 2012). However, given that the boom of mobile gaming in Korea started with smartphones, which have a large-capacity touch screen, keyboard, and applications, *mobile gaming* should be used to refer to games and gaming culture specifically using cell phone platforms, particularly smartphones, while excluding other portable devices such as Nintendo and Sony PlayStation handhelds. This does not mean that I disregard the significance of these portable devices for mobile games, but I focus on mobile gaming on smartphones in order to fully define the role of smartphones in the boom of mobile games in Korea.

On the other hand, we need to understand that the recent boom in mobile gaming is not discrete from previous developments, but is part of the transition caused by the growth of the smartphone era, as the advancement of online games has relied on the development of high-speed Internet and the PC bang (i.e., rooms in Korea where players pay to play online games; Jin 2010; D. Lee 2012; Jin et al. 2015). In fact, people's "leisure/entertainment styles have transformed with the advent of ubiquitous smartphone use in everyday Korean life. Instead of watching popular dramas (tŭrama, Korean television dramas) on television or other portable devices, users have come to watch these programs on their smartphones in public places, including the bus, train, and subway, as well as school. Smartphones have particularly become the major platform of choice for mobile gaming. In Korea, online gaming has been the most influential on both youth culture and the game market, but gaming has witnessed a gradual change toward this trend in very recent years" (Jin et al. 2015, 417). Due to several popular mobile games in the early stage of the smartphone era, including Anipang and Candypang (Kaendip'ang), previously dormant gamer groups, such as many females or people in their thirties or older, started to play video games. The notion of game culture is frequently used in the field of adolescent video game research to refer to the distinctive ways in which youths around the world embed video games in their everyday lives. However, mobile gaming on the smartphone has expanded digital game culture, from mainly youth oriented to an all-age-group-driven culture (Abeele 2016).

Thanks to the touch screen and mobility of smartphones, "Mobile game developers produced casual games, which are characterized as a mode of engagement that requires only sporadic attention up to a threshold of around five minutes" (Richardson 2012, 143), and distinguished them from online games by their simplicity for game users. Several mobile games, such as Anipang—a social-type puzzle game launched in 2012—and Candypang harvested abundant downloads during the second half of 2012, which opened the full-scale mobile game era. Inspired by the phenomenal success of Anipang, Sunday Toz, its developer, created Anipang 2 in January 2014, which became the second most popular mobile game, behind only Monster Taming (a role-playing game) by Netmarble. In May 2014, Anipang 2's monthly active users numbered 4.7 million (Kim Sang-yun 2014). Anipang games on smartphones generated a nationwide mobile game boom.

The consequences of the growth of mobile gaming are noticeable. As a reflection of the phenomenal growth of mobile games in recent years, for example, Blade for Kakao won the nation's most prestigious game prize during the 2014 Game Awards Ceremony—the first time a mobile game had earned the top spot in the award's 19-year history. This can be seen as further proof of mobile games' fast-growing presence in the Korean game market. "Blade for Kakao is a 3D mobile action role-playing game developed by Action Square and published by Four Thirty Three;" it was launched in April 2014. "The publisher revealed that it recorded five million downloads and accrued 90 billion won ($81 million) in sales over six months in the domestic market. This sales figure is the largest ever for a single mobile game over a half-year period. On the back of the popularity of the title, Four Thirty Three secured an estimated 120-billion-won investment for the Asian market from Chinese game giant Tencent and Japan's Line" (S. Yoon 2014).

As mobile games have become a primary game form for many Koreans, several mobile game corporations have rapidly expanded their investment in the development of new mobile games in the era of smartphones. For example, Com2uS and Gamevil, the two major mobile game corporations in Korea, became one company in October 2013. Both companies had battled for the local Korean mobile gaming market; however, Gamevil acquired Com2uS for $65 million—a 21% stake in its longtime competitor (Cutler 2013). The major reason for this acquisition was to deal with the increasing role of platforms as outlets. As discussed, in the midst of the rapid growth of the smartphone, U.S.-based operating systems, both Android and iOS, have dominated the market, and these two operating systems have flattened the mobile gaming market. At the same time, powerful new apps like Kakao Talk are wedging themselves between the Android platform and domestic game developers, lessening the power of local studios (Cutler 2013). Com2uS and Gamevil expected that they might be able to do better together in the face of current industry shifts. Meanwhile, NCsoft and Netmarble Games, two of the country's major game companies, established a stock-swap alliance in February 2015. NCsoft needed to have Netmarble in order to successfully enter the mobile game market, and Netmarble wanted to have NCsoft's global online game intellectual properties, which made a cross-marketing initiative possible (S. Yoon 2015a). Unlike several years ago, when online gaming was the major interest, the Korean video game industry has rapidly turned its eyes to mobile gaming, which has consequently created large game corporations and blockbuster-style mobile role-playing games, which will be detailed later.

Sociocultural Dimensions of the Popularity of Mobile Games in Korea

There are several sociocultural dimensions and government policies pertinent to the rapid growth of mobile games. While it is not my intention to analyze government policy, it is certain that one of the key elements in the development of mobile gaming has been favorable government policies. The Korean government believes that the mobile game sector is the next growth engine in the digital economy in conjunction with smartphones, and it has supported the mobile game industry. In particular, deregulation has promoted the recent growth of the mobile gaming sector. Korean game companies had suffered under some rather heavy-handed regulations; however, the so-called Open Market Law was passed by the National Assembly in March 2011, relaxing censorship (H. Kim 2011). The Open Market Law implies that Korea's Game Ratings Board no longer assesses games made available on feature phones, smartphones, and tablet computers, with the exception of titles featuring gambling and adult content. "In the past, all games required a review by the Game Ratings Board before they could be made available on app stores. Google and Apple were unsatisfied with those terms and shut down their local app stores until Korea passed its new law" (Caoli 2011).

Under this circumstance, several sociocultural elements have played a key role. The Korean sociocultural milieu has become a primary driver for the development of mobile gaming, because people as consumers use and enjoy mobile games on their smartphones. As Korea's online gaming has grown based on the country's unique sociocultural dimensions, such as the rapid growth of broadband, the increasing role of PC bang, mass play culture in a group-oriented society, as well as the development of e-sport (electronic sport) as a main part of youth culture (Jin 2010), Korea's distinctive sociocultural factors have become major elements for the recent advancement of mobile games. Among these, the optimized infrastructure for the development of mobile gaming, such as the growth of smartphones and relevant apps, including Kakao Talk, and the rapid growth of 4G LTE, as well as the rapid penetration of wireless broadband services, have become major contributors. The unique commuting patterns in major metropolitan areas, such as Seoul and Pusan, are an impetus to the growth of mobile games. Unlike online gaming, where the majority of users are males in their teens and twenties, the increasing visibility of female players and middle-aged players on the social grid and a peer-driven cultural setting have also become primary reasons for growth in the breadth of the gaming community, although male users also enjoy mobile games (Jin et al. 2015). These sociocultural dimensions have shaped and influenced Koreans' daily activities and their cultural lives, resulting in the growth of mobile games in the context of the Korean smartphone space.

The Commuting Pattern in Korea's Urban Setting

Korea's unique commuting pattern in a dense, mostly urban environment plays an important role in the country's widespread consumption of mobile games. The Korean subway and train system is vast and heavily used. Of Korea's population of 50 million, half live in Seoul and its metro area (Acuna 2013). Ridership on the Seoul subway system increased about 14.2% from 2008 to 2013, from 2.29 billion rides to 2.61 billion rides in 2013 (Seoul Metro 2014). This ranks the Seoul subway system the second busiest in the world, behind only Tokyo. Korea's modern, LTE and Wi-Fi-equipped subway cars are comfortable places for gaming. "Daily commutes can be long and uncomfortable. Children and professionals escape the shoulder-to-shoulder morning rush, not by reading books, but by gaming and chatting with friends on Kakao Talk" (Acuna 2013). As many people living in the Seoul metro area commute to and from Seoul via the subway, mobile gaming is one of the best ways for the commuters to kill time.

In response to this distinctive commuting pattern, mobile service providers have started to sell Subway Free Data Plans, which are value-added plans, since June 2014. The customers who already have their LTE plan can pay between 5,000 and 9,000 won to access data up to 2GB per day if they want to enjoy several services while commuting. The subway system in

Fig. 11. Mobile Game Advertising on the Gangnam Station's Subway Screen Door. (Photograph by Dal Yong Jin, December 2014, when DOTA Legend [the original title is Heroes Charge], a mobile game developed by Lilith Games and published by Longtu Games, was released and played in Korea.)

Seoul naturally has been targeted by mobile game corporations for their board advertisements (see fig. 11). The subway system in the Seoul metro area, therefore, is a place to experience all kinds of mobile culture, from game culture to commercial culture.

During my interviews, a female researcher at Hyundai Motors (age 26) who likes to play Angry Birds said, "Angry Birds is usually for killing time. I play Angry Birds when I am bored at home or waiting for the subway or an appointment." A male graduate student (25) stated, "I use Kakao Talk a lot on my smartphone, and I also make many calls. I do Kakao Talk for killing time on my way to and from work, and I do mobile games, such as Plants vs. Zombies, Paper Zombies, Angry Birds, and Temple Run." The mobile game craze, if anything, is more "pronounced in the urban setting that forms the crowded life of Seoul and other cities. Smartphones and mobile gaming certainly represent mobility and transience in urban places" (Jin et al. 2015, 421).

In tandem with the unique commuting pattern in the urban setting, another primary characteristic feature of the environment of mobile gaming is mobility. While online game players sit facing a PC for several hours at home or at a PC bang, mobile games challenge traditional definitions of play and force us to look at city spaces as playful spaces, including subway and buses (Silva and Hjorth 2009). In other words, "Playful spaces are mostly urban spaces and are produced by the mobility and interactions of people who inhabit these spaces" (Silva and Hjorth 2009, 604). As Lemos (2011, 289) points out, "Mobility is especially deterritorialization—virtual, physical, or imaginary—and transportation and communication technologies are a way to reinforce these mobilities. By playing mobile games on the subway and bus with smartphones and wireless networks, mobile games manifest both physical and imaginary mobilities." During the interviews, many respondents said that they played mobile games everywhere when they had spare time. What they emphasized was the mobility of such games, naming simple and brief mobile games. "Mobile games, based on mobility and play, transform public spaces into playful spaces" (Silva and Hjorth 2009, 622), as online players cannot walk or bring a desktop computer to the city.

Community-Oriented Cultural Milieu

Korean video games, both online and mobile, rely heavily on a community-based ethic deeply embedded in Korean culture. As several successful online games including Lineage exemplify, its attraction is a community-based game structure, which has been a very important feature for Korean online gamers. Lineage's developer and publisher (NCsoft) claims that players have more fun by joining hunting parties and forming allegiances, heightening their abilities to fight foes and monsters. This gives gamers the sense of there being a common goal and the chance to socialize with other players (S. J. You 2003). The mass play culture of Korean youth has become one of the major contributing elements in the growth of online gaming because they like to play with friends (clans and/or guilds in games), and NCsoft certainly utilized this unique Korean culture in producing Lineage (Jin 2010).

This "community-based social environment has also become significant for mobile games, because mobile gaming has tapped into the country's tendency to emphasize community formation" (Jin et al. 2015, 419). The community-oriented online culture has been embedded in people's mentalities and has correspondingly influenced mobile game culture. In other words, Korean mobile users take it for granted that they will play mobile games with friends, which are slightly different because of the platform. The communal feeling is unique, because mobile gaming has typically been associated with solitary game play in other countries (Jin et al. 2015).

According to a survey conducted by Korean consulting firm NAS Media (2015), Korean mobile gamers start games primarily because of their social relationships. Mobile gamers install and play mobile games when their friends recommend specific games (51.2%), when the game is top-ranked (42.2%), and when they have been affected by advertising (27.6%). This implies that Korean mobile gamers are naturally and passively exposed to mobile games in the community-driven society.

Yi Jŏng-ung, CEO of Sunday Toz, explained that the single major reason for the success of the Anipang series is its heart system (Im 2013). In order to play an Anipang game, people need at least one heart, and friends can give a heart to their friends as a present, which is an example of a community-based mobile game. Im explains that one of the most significant means for the growth of mobile games is the network of players, and by sending hearts, Anipang continues to increase its number of users. Anipang players also send and receive hearts so that they increase their friend's score. A seemingly individualized mobile game is actually a very well-networked game because friends compete with each other, in that they share their scores and therefore compete to be the best player in their community network.

This kind of community-based success story has greatly influenced other mobile games. For example, Monster Taming, made by Netmarble and named in 2014 the mobile game of the year, utilized the same format by sending and receiving a key to play and to enhance friendship.

When my first round of interviews was conducted at the end of 2012, several respondents indeed replied that they started playing mobile games, in particular, Anipang and Candypang, because their friends and colleagues recommended the games and sent hearts to invite them. In the case of online gaming, many Korean youth play together at a PC bang; however, in mobile gaming, users socialize with each other without physically gathering by inviting their friends from anywhere. One female businesswoman in her early twenties said:

> I enjoy mobile games being able to freely connect to the Internet so that I can have much more fun from playing with others (friends and random people from a social network). For example, I like minigames that allow one playing against the other, visiting each

other's village or farm in the game and so forth. Sometimes it generates a sort of community.

During an interview conducted in July 2015, a male college student (21)
said that his favorite mobile game is Clash of Clans because all of his
friends play and it is addictive. He also likes it because "it does not require
a lot of time to understand how to play," it allows him to socialize with his
old friends because the game is connected to Facebook, and a lot of his
high school friends started to play because of that connection. One of the
interesting features of community-based mobile gaming is the ranking
system, as clearly indicated in Anipang games. For example, another male
college student in his early twenties stated,

> I like Candy Crush most. I often play it on the subway, and I think
> it is a good choice for killing time. However, I also enjoy this kind
> of mobile game because of the ranking system among my commu-
> nity. In this way, the game makes the connection between friends
> and family members.

On its home page, Candypang states, "Receive hearts by inviting Kakao
Talk friends, enjoy with friends, and challenge the best scores" (WeMade
2012). In regard to this network of friends, Rainie and Wellman (2012, 127)
point out that "people use the Internet and mobile phones to keep in
touch, to arrange get-togethers, and to follow up after they meet." As online gaming has long been a virtual space for many Koreans to socialize,
mobile gaming on smartphones has suddenly become a unique tool for
many students and businesspersons to connect to each other while enjoying their pastime games.

Rapid Adaptation of Kakao Talk in Digital Technologies

Korea is well known for its swift adaptation of new technologies, from
broadband services to PC bang to smartphone apps. While people
around the world are keen about cutting-edge technologies, Koreans' acceptance of new media-related technologies has not been gradual, but
once and for all. As discussed in previous chapters, Korea was late in its
Internet revolution; however, since the middle and late 1990s, Korea did
not take long to adapt new media technologies, which have resulted in
the phenomenal growth of the smartphone and relevant apps. This certainly indicates that people as major users are taking a key role in the

growth of new media technologies. The growth of mobile gaming would not have been possible without consumers rapidly accepting and adopting new technology, in particular Kakao Talk. The cultural mentality embedded in most Koreans has caused the increasing popularity of mobile gaming with the swift growth of smartphones and new distribution platforms, including Kakao Talk.

As discussed in chapters 1 and 8, Kakao Talk began as a mobile messenger service but has transformed itself into a platform for the distribution of diverse third-party apps and content, including mobile games that users can play with their friends through the messaging platform. As Kakao Talk Inc. itself states, mobile game applications struggled to attract one million downloads until the very early 2010s. "Kakao's game platform completely reversed this trend, and gave birth to eight games that recorded more than 10 million downloads as of July 2011" (Russell 2013). Partially due to Kakao Talk and, of course, other apps, Korea recorded a 759% increase—the highest growth in the world—in the mobile app market in 2013, followed by China (280%) and Japan (245%) (Dredge 2013). Kakao Talk has substantially played a significant role in generating the nationwide drive in the usage of smartphones because people, in particular, youth, have to get smartphones to enjoy the diverse services Kakao Talk provides (Jin et al. 2015).

As a female game developer (24) who created a mobile game with her college friends explained:

> I use Kakao Talk group chatting with friends for planning a get-together or a trip. I also use Kakao Talk group chatting with co-workers from my current start-up company. I have a group chatting room with my family members and I think it promotes the family spirit.

Kakao Talk, which offers free in-app texting and voice calls as well as a platform for a world of mobile services, has become one of the most significant drivers behind the transformation of the game market toward smartphone-based mobile gaming. Kakao, along with its partner studios, offers these games on the chat app. "Kakao Talk has become the greatest single source driving mobile-game downloads. No other source for mobile-game discovery even comes close to driving as many downloads. As of May 2014, Kakao Talk has over 230 partners that it works with, and it has promoted more than 450 games on its service" (Grubb 2014).

However, it is also important to acknowledge the recent weakness of

Kakao Talk as the leading platform, because many mobile corporations establish and/or find their own platforms in order to avoid higher service fees that these game corporations pay to Kakao Talk. Several platforms comparable to Kakao Talk, such as With Band and With Africa TV, as well as in-house game platforms, including Hive by Com2uS and Chef, operated by Identity Mobile, are representative alternatives to Kakao Talk. In addition, Naver—the largest Internet portal in Korea—developed With Naver, which is relatively successful (Ministry of Culture, Sports and Tourism 2015, 32). Thus, in very recent years, platforms have competed with each other to attract new mobile games and, therefore, mobile game players. It is premature to draw conclusions about the success of these new platforms in the near future; however, it is certain that Kakao Talk needs to work with mobile game corporations by lowering its service fees, because the service fee is one of the major reasons for the emergence of new game platforms. Otherwise, it may lose its leading position as the major mobile game platform. Only a few years ago, blockbuster hits of mobile games in Korea were almost always Kakao Talk games; however, this is not the case anymore, because foreign-based apps, including Google Play, have also developed their own strategies to penetrate the Korean game market.

Transformation toward Mid-core Role-Playing Game Culture

The increasing visibility of mobile games in Korea shows a new form of transformation—from a female-driven casual game stage to a middle-aged users-oriented mobile role-playing game stage, as mobile game genres have shifted from casual games to RPGs. Up until several years ago, casual games were the major genres targeting female game players. However, what famous male actors represent these days are role-playing games, which have been major online games on the smartphone.

To begin with, as in many other countries, game users analyzed by gender have shown a very interesting trend in the Korean game market. When the iPhone was introduced in Korea in 2007, both male and female game users enjoyed online games. Back then, interestingly, female users (83.4%) played online games most, about 10 percentage points higher than male users (73.3%). Although there was no significant difference between male (33.5%) and female users (32.6%) in enjoying role-playing online games as their most popular game, female players enjoyed casual online games (18.3%) more than male players (9%) as their second most popular game genre (Ministry of Culture and Tourism 2007, 333–35).

The situation was not changed until 2010 right after Korea received iPhones and started to sell its own smartphones. When mobile games were not popular yet, male players (72.4%) and female players (73.5%) were still enjoying online games. Unlike in previous years, though, female players increased their preference for mobile games, from only 4.7% in 2007 to 13.2% in 2010. The picture in the video game market has rapidly changed since then.

By gender, in 2013, male players still enjoyed online games most (51%), although the proportion had dramatically dropped. However, female players played mobile games most (45.9%), followed by online games (25.9%) (Ministry of Culture, Sports and Tourism 2013, 411). Female players have certainly changed their preference for mobile games, primarily because they like casual games on the smartphone. For female players, easy control, free access unrestricted by time and place, and adorable game characters are crucial, and mobile games satisfy all of these conditions (*Korea Times* 2012).

Several scholars argue that female players primarily play mobile games, while male players enjoy online games. As mobile phones themselves also target female users (Shade 2007), many mobile game designers develop female-driven mobile games.[1] Hye Ryoung Ok (2011, 334), for example, argues that "many women see online and offline game worlds as constructions of masculine space and feel social restraints or societal pressure in navigating these worlds"; therefore, female players feel comfortable with mobile games. As Vanderhoef (2013) also points out, hard-core games become the dominant masculine genre, while casual games become the feminine, and the terms "*mobile* or *social* are now more widely associated with casual games"; likewise, smartphones and tablets are the current popular devices to play these games on.

However, as Korea's development of mobile gaming in the 2010s demonstrates, this kind of simple categorization cannot explain a rather complicated video game culture. As discussed above, unlike some perceptions, Korean female players enjoyed online games greatly, even more than male players, including MMORPGs (Ministry of Culture and Tourism 2007). On the other hand, male players have also rapidly moved to mobile games, virtually matching female players in their preferred games. One of the major reasons is the rapid growth of mobile role-playing games, which has created a second stage of development—growth driven by middle-aged players—in recent years.

In 2013, for example, while online games were mainly played by teens and people in their early twenties (especially those under 24 years old), mobile games were played by middle-aged groups in their thirties and forties; however, many in their upper twenties (over 25) also enjoyed mobile games more than online games (Ministry of Culture, Sports and Tourism 2013, 412). These age groups are mostly working-class people in diverse companies who need to commute via subways and buses. Since they do not have enough time to play time-consuming online games, they used to play casual games on their smartphones. However, as mobile role-playing games have become popular, male video game players are shifting their preferences from online games to mobile games. In 2015, the two largest game corporations in Korea, Nexon and Netmarble, released their new mobile RPG: Heroes of Incredible Tales (HIT) by Nexon and IDEA by Netmarble. Previously mobile games were developed within a very short time with a low budget; however, mobile game corporations spend a lot of time and money to develop new mobile RPGs. For example, Netmarble took three years to develop IDEA, and the production cost was as much as $10 million, one of the highest budgets in Korea's mobile game history (T. Yi 2015).

As a reflection of the changing demographics in the mobile game sector in which people in their thirties and forties, including male businessmen, have become the biggest consumer group, Korea's game developers have developed their distinctive ad-marketing skills. They have used middle-aged Korean actors with top-class popularity because of their machismo image to target potential buyers in their thirties and forties, which is a unique approach in Korea. Several top-tier male actors, including Cha Seung-won (Ch'a Sǔng-wǒn), Hwang Jung-min (Hwang Chǒng-min), Jung Woo-sung (Chǒng U-sǒng), Lee Jung-jae (Yi Chǒng-jae), and Ha Jung-woo (Ha Chǒng-u), have appeared in ads for mobile role-playing games that launched in 2015. Cha, 45, was the model for Raven, and Ha, 37, became the face of ChronoBlade. Hwang, 45, Lee, 42, and Jung, 42, appeared on behalf of Asker, Ghost (Kosǔtǔ) with Rocket, and Nantu, respectively. Hollywood actor Lee Byung-hun (Yi Pyǒng-hǒn), 45, was featured for IDEA and Jang Dong-gun (Chang Tong-gǒn), 43, for MU Origin. Of course, mobile game corporations and ad agencies have selected these top actors in their thirties and forties because they can attract both male and female consumers (*Korea Times* 2015).

Mobile gaming is not limited to its own industry, but has expanded its convergence with the broadcasting and advertising industries. This implies that mobile gaming is no longer a pastime for dedicated gamers only, but also for the general public, as television ads become a new means to advertise mobile games (Ministry of Culture, Sports and Tourism 2015, 36). There has been no other particular cultural industry where middle-aged adults, in particular male users, have become major players in Korean cultural industries.

According to a survey conducted by KT Economic Institute and Nielsen Korea in September 2014, Koreans spend an average of three hours and 40 minutes per day on their smartphones. People in their thirties and forties mostly play games, while the young fritter away their time messaging. Those in their thirties and forties play games most often, at 61 minutes for users in their thirties and 52 minutes for those in their forties, and in general, Koreans over 30 use smartphones for games most, while teens and people in their twenties use smartphones for telephony, followed by games. These data certainly prove that mobile games are enjoyed by middle-aged players more than online games. The issue of changing demographics explains the current trend: "The first generation of people who played video games as children are now well into their late thirties and early forties, have less time on their hands than they used to, but are looking for video game experiences that work for them today" (Juul 2010, 147).

Female players are not the only groups in the new mobile game space anymore. Previously, for female gamers, mobile games provided spatial freedom from the male-dominant social order in conventional game spaces (Jeon 2007, as cited in Ok 2011, 334). However, unlike the U.S., where female players are still dominant in the mobile game sector, in Korea male and female players are almost equally present, mainly because of a much higher trajectory for RPG and action-adventure games, with those games grabbing 43% of the market in Korea (Usher 2014). Female players in the realm of mobile gaming were major users at least for a while; however, the milieu of mobile gaming in Korea has rapidly changed.

As many male players have continued to play online games because they relatively enjoy hard core role-playing games, mobile game corporations have rapidly turned their strategies toward develop RPGs on smartphones. Therefore, mid-core gaming is a segment that is attracting increasing attention from investors and developers. Games that are in between what we traditionally call "casual" and "hard core" games have always been around, and combining an immersive experience and casual gameplay is what mid-core gaming is about. From a game player's perspective, mid-core gaming is comprised of players looking for a more in-depth experience than a casual game provides, yet one that is not as time-consuming as a core game. The potential for mid-core games is illuminated by the fact that former core players, as they start families and careers, no longer have as much time to devote to playing; the pool of potential players also includes casual gamers who desire a more immersive experience (Warman 2012). Thus, it is quite challenging to predict the role of gender in the mobile game market, although it is clear that female players and male players, in particular, those over 30, will be major users of the mobile game sector, in which female players were dominant. As Jesper Juul (2010) and Ingrid Richardson (2012, 143) suggest, "The stereotypes of the hard-core and casual gamer over-simplify the often complex and variable modalities of play, and . . . the recent proliferation and popularity of casual games reach across many demographics." On top of that, mobile game designers are developing mid-core mobile role-playing games targeting both genders in all age groups. Therefore, gender-based genre classifications in mobile games cannot be narrowly drawn in our contemporary game world, at least in contemporary Korea.

Transformation toward Mobile Game Culture

As mobile games have replaced online gaming in recent years with the phenomenal growth of the use of smartphones, policymakers, game designers, corporations, and game players ponder the future of these two major video game platforms in Korea. What they mull over is whether people can see the changes in people's game habits, from traditionally strong online gaming to emerging mobile gaming culture. Since many Koreans, not only middle-aged game players but also teens and players in their early twenties, are shifting their favorite games from online to mobile games, this is the logical question to be considered.

While there are several significant areas to be discussed in order to determine the potential transformation of Korea's video game market, in this final section I want to discuss the continuity and change of video game culture by addressing two major issues. One is the identifying characteristics representing online gaming and mobile gaming, which are "hard-core online gaming" and "casual mobile gaming," respectively. The other is the role of game corporations.

On the one hand, in the realm of video gaming, MMORPG, as a representative game genre, has been classified as hard core. "There is an identifiable stereotype of a hardcore player who has a preference for science fiction, zombies, and fantasy fictions, has played a large number of video games, will invest large amounts of time and resources toward playing video games, and enjoys difficult games" (Juul 2010, 8). In contrast to this, mobile games are primarily casual games. "Casual mobile gaming is often characterized as a mode of engagement that requires only sporadic attention up to a threshold of around five minutes, hence the popular notion that casual games are the mobile phone's predominant game genre, and the labeling of casual gamers, who play at most for five minutes at a time and at irregular intervals, as a key market in the mobile game industry" (Richardson 2012, 143). As Jesper Juul (2010, 8) points out,

> Casual games apparently reach new players, and the new players they reach are often called *casual players* . . . . The concepts of casual players and casual games became popular around the year 2000 as contrasts to more traditional video games, now called *hardcore* games, and the hardcore players who play them. Casual players are usually described as entirely different creatures from hardcore players. . . .
>
> The *stereotype of a casual player* is the inverted image of the hardcore player: this player has a preference for positive and pleasant fictions, has played few video games, is willing to commit little time and few resources toward playing video games, and dislikes difficult games.

The industry classification of "casual games" encompasses several genres—digital puzzle, word, and card games such as Candy Crush Saga, Words with Friends, and Solitaire, and also time management and social games such as Diner Dash and Farmville. These very different games share some basic similarities: they have "simple graphics and mechanics, they are usually browser or app-based, and they are free or cost very little to play. Most importantly though, casual games are designed to be played in short bursts of five to ten minutes and then set aside" (Anable 2013).

Arguably, many people prefer mobile gaming to online gaming because of its usability—the players can play games without understanding how to play, as a pastime in Korea's unique sociocultural milieu. In this regard, during our interview, a female businesswoman (24) stated, "I prefer games with excellent/pretty graphics and an interesting story." She plays such games on a PC because such games usually can be run only on a PC, although she plays smartphone games because games on PC are not portable.

On the other hand, game users' changing cultural patterns have also influenced the game industries. As discussed, with the rapid growth of large-screen smartphones, mobile game developers, including Netmarble, Nexon, and WeMade, have developed role-playing games in addition to traditional casual games. As Juul (2010, 148) points out, "The economics of video game development are already quite uneconomical because of soaring development cost for hard core role-playing games. Due to their smaller scope, casual games are generally cheaper to develop than the larger hardcore games." However, due to the increasing use of the smartphone in people's daily activities, mobile game corporations have invested in creating mid-core mobile role-playing games. Mobile game corporations work with foreign game corporations in order to secure money and games. For example, Tencent, China's largest Internet company by market capitalization, stepped up its commitment in mobile gaming by announcing its plan to buy 28% of the stakes in Korea's CJ Games for $500 million. Tencent and Line Corp. also jointly invested $100 million in Korean mobile game designer 4:33 Creative Lab in November 2014 (Bischoff 2014). Because of the rapid growth of mobile gaming on smartphones, many foreign game corporations have switched their investment from online to mobile games, in particular mobile RPGs, in recent years.

Foreign mobile developers and publishers have also targeted the Korean mobile game market, because it is a test bed for global game corporations. Finland-based Supercell, for example, massively promoted its global hit mobile game Clash of Clans in 2014 and 2015 to make the game the most popular mobile game in Korea. Supercell announced that it would use $20 million for advertising in Korea when it released Clash of Clans in June 2014, and it paid off (Shin 2014). The game became the top mobile game on Google Play in October 2014, a position it held until the first week of January 2015. As Kim Sung-gon (Kim Sŏng-gon), an official from the Korea Internet and Digital Entertainment Association said, "If a company can succeed in the most wired and highly competitive domestic market, it is highly likely to have good performance in other countries" (Shin 2014).

These recent developments ask us to reconsider our perceptions of mobile games as consisting of predominantly female-oriented casual games (Richardson 2012), and the rise of casual games shifts the perspective on whether a game fits into the life of a player (Juul 2015). Since the major characteristics of online and mobile games are not the same, it is not easy to grasp the entire shift of online gamers to mobile games. It is safe to say, however, that online gaming and mobile gaming go hand in hand in Korea's distinctive sociocultural video game setting, which has been influenced by online games for almost two decades.

Conclusion

This chapter has analyzed the growth of mobile gaming in the era of the smartphone. Since Korea has been known for its online gaming since the early 21st century, mobile gaming in Korea has been relatively a late comer, but it has experienced remarkable growth in terms of its market share in the domestic market in tandem with the growth of smartphone use. While the transition of mobile phones into smartphones has occurred alongside the rise of mobile gaming, the recent breakthroughs in smartphone technologies have especially contributed to the growth of mobile role-playing games. Although it is not yet comparable to the popularity of online gaming, mobile gaming has suddenly become a significant popular cultural practice for many Koreans, as the size of the smartphone's screen and apps rapidly advance. Korea has developed mobile gaming based on several distinctive characteristics, such as a culture that emphasizes community, urban life, and quick adaptation of new media, in this case smartphones and relevant apps pertinent to mobile games. In other words, Korean mobile games cannot succeed separately, but as part of a particular sociocultural environment.

As Marshall McLuhan (1994) pointed out, technological innovations do not necessarily introduce absolutely new elements into human society, but may nonetheless accelerate and enlarge the scale of previous human functions, which in turn create new lifestyles (work and leisure). In no other country has this acceleration been more concentrated and apparent than Korea. Mobile gaming, as our case has shown, does not develop independently, but as part of a particular sociocultural milieu. It has driven the transformation of Korea's networked society through a mutual social shaping of technology use (Jin et al. 2015, 427).

Overall, while time is still needed to determine the coming trend in the Korean mobile game market, it is certain that a new chapter is opening in the fastest-growing smartphone and app businesses around the world. Korea has experienced a transformation of its digital game industries, mainly from being online game-driven to a situation saturated by mobile gaming, which has also transformed youth culture as embedded in digital games.

8

Reimagining Smartphones

Kakao Talk and Youth Culture

In smartland Korea, smartphone apps have provided convenient platforms for local ICT users. Kakao Talk especially plays a key role in shifting people's daily activities because of its tremendous impact on smartphone users, in particular young users. As previously discussed, since people's daily activities rely on the smartphone rather than the Internet, primarily because of its functionality and mobility, as well as playfulness, Korean smartphone customers use Kakao Talk as their most significant cutting-edge app. Kakao Talk has rapidly become part of everyday social actions. Kakao Talk has effectively replaced text messages, and many Koreans use their smartphone mainly because of Kakao Talk. People in Korea often say, "Ka-talk (k'at'ok) me" instead of "E-mail me" or "Send me a message." Released in 2010 as a free download onto smartphones, Kakao Talk became an instant hit. After quickly overtaking Facebook and Twitter, Kakao Talk became Korea's number one social networking service and a provider of wildly popular mobile games (E. Choi 2013). Smartphones allow users to connect with other people and to share information, news, and content (Humphreys 2013), and for many Koreans, Kakao Talk on the smartphone is the most important social medium—one cannot live without it. Kakao Talk indeed "introduced to the world the big insight that messaging apps can be far more than chat. They're platforms for selling all kinds of lucrative services such as games and virtual goods such as coupons and stickers" (Mac 2015), and therefore, it has dramatically influenced "the sphere of everyday life and people's communicative practices, personal relationships and social experience" (Linke 2013, 33). As discussed in chapter 7, Kakao Talk has been one of the key driving forces in the growth of mobile games. Kakao Talk offers a vivid example of how smartphone evolution is reimagined in a local context. In particular, young Koreans' engagement with Kakao Talk and related mobile apps suggests that smartphones have become symbolic and material resources for their urban lifestyle.

The emergence of Kakao Talk as a platform asks us to explore how Kakao Talk is integrated into the socioeconomic and cultural landscape in a particular local context, and how young ICT users engage in sociocultural activities. This chapter examines a localized media environment that arose with the smartphone and its apps, with particular reference to young Koreans' engagement with the local app platform Kakao Talk. It analyzes the "Kakao talkscape" as a form of mediascape embedded in and constructed by specific sociocultural circumstances in Korea. In this respect, it pays attention to the ways in which users have engaged with the socioeconomic processes behind the rapid diffusion of smartphones and Kakao Talk.

Understanding the Smartphone in Youth Culture

With the groundbreaking rise of smartphone-mediated communication among youth across the globe, several scholars have conducted empirically grounded cultural analyses of "smartphone fever." Existing research on young people's engagement in mobile technologies has contributed to the rapidly emerging field of smartphone studies. Until a few years ago, the lack of an empirical basis in the study of smartphones as a new technology led to a failure to explore the smartphone as a medium of convergence and divergence (Watkins et al. 2012). As Jones et al. (2013, 3) point out, "The lineage of mobile media and communication can be traced back to studies of the mobile telephone. The first analyses came in the early 1990s as mobile telephony began to diffuse into society." However, analysis of smartphones started less than 10 years ago, given their short history, and the smartphone offers a media platform in which various practices— communication and play in particular—converge, creating social space in which locally diversified production, consumption, and imagination emerge, resulting in the growth of scholarship on the smartphone.

In pursuit of a critical-cultural understanding of the smartphone and Kakao Talk, it is crucial to understand the nature of each, and, in particular, it is pertinent to consider the smartphone not as a single form of technology, but as the convergence of different media technologies. It is a gateway to several goods, services, and cultural practices, especially leading to the growth of hundreds of thousands of applications (Christensen and Prax 2012). Thus, it would be difficult to explore social and cultural meanings of the smartphone if we relied on a single dimension of the technology. In this respect, as Wilson et al. (2011) have suggested, we need to consider the smartphone as both communication and gaming technology, by drawing upon conversations between "mobile studies" and "game studies." This framework will be incisive for understanding the cultural significance of smartphones for young users, in that it addresses their enhanced communicative sociality ("mobile studies") and playful performance and identity work as a narrative form ("game studies").

First, smartphones' and Kakao Talk's communicative roles in youth culture can be examined through mobile studies that have vigorously dealt with mobile media technology as an individualizing and socializing cultural form. From this perspective, the smartphone can be seen as a means by which young people express their identities, socialize with others, and participate in the public sphere. On the one hand, aesthetic aspects of mobile phones have been analyzed as a signal of young people's pursuit of expressive and individualized lifestyles, which also confront the dominant social order (Ling and Yttri 2002; McVeigh 2003; Ito and Okabe 2005; Katz and Sugiyama 2006; Hjorth 2009; Linke 2013). In other words, it is argued that the mobile-mediated, highly individualized modes of communications among youth have symbolic meanings that challenge the dominant cultural norm. The mobility and immediacy of mobile phones can create a space for constant contact, which eventually contributes to "an increase in social interactions among spatially distributed family members and acquaintances in daily life. However, smartphones allow people to extend their mobile communicability to the online social space, reconfiguring the forms and scopes of mobile communication practices and thereby the conditions of people's sociality" (D. Lee 2013, 271).

On the other hand, the mobile phone facilitates sociality, enhances social capital, and mobilizes political activism (Ling and Yttri 2002; Hjorth 2009; Ok 2011; Qui and Kim 2010). It strengthens the existing micronetwork (K. Yoon 2003) and creates new weak ties through which young people can acquire social and cultural resources (Yuan 2012). Furthermore, it stimulates young people's participation in political activism, as exemplified by young Koreans' massive antigovernment protests throughout the early years of the century (Ok 2011; Qiu and Kim 2010). Mobile studies have claimed that young people, increasingly equipped with mobile technologies, form highly flexible affiliations with each other, which can be constructed "smart mobs" (Rheingold 2002). "This perspective challenges the collective and rigid notion of youth as a subcultural unity, with the realization that an increasing number of ordinary young people get involved in 'play, parody, humor, wit and caricature to express their feelings' via mobile technologies" (Ok 2011, 328).

Meanwhile, the playful aspects of smartphone and Kakao Talk use among the youth can be analyzed through game studies. In the past two decades, cultural studies of gaming have explored games' textual features and later, their ludic aspects. While earlier game studies, or "narratology," appropriated narrative analysis in order to "read off" games as texts (e.g., Wolf 2002), the rise of mobile gaming, role-playing games (RPGs), and "casual" gamers has made the narratology approach outdated. Thus, resonating with active audience studies drawing upon ethnographic methods, another approach, "ludology," has emerged to explore gaming as performance and play (Juul 2005). As the forms and interactivity of gaming have become more diversified, mobile, and casual than ever, it appears necessary to identify the players and how they do it. In particular, the ludic experiences with RPGs are highly suggestive in understanding smartphone-mediated youth culture, in that the player's collaboration with others, identification with different characters, and engagement with social space through storytelling in RPGs (Call et al. 2012) are easily applicable to smartphone users. Smartphones have enhanced the mobility and playfulness of gaming, while enabling a wide range of genres such as multiplayer, urban, mobile location-based, and hybrid reality games (Silva and Hjorth 2009). The smartphone mediascape constitutes a gaming milieu where online yet sedentary gaming practices are increasingly being transformed to mobile and casual ones among youth living around big cities (Jin et al. 2015).

An articulation of mobile studies and game studies may contribute to exploring the implications of smartphones and Kakao Talk in youth culture. However, not much research has delved into the manner in which communicative and playful dimensions of smartphones are articulated with each other. The lack of empirical studies seems to entail another problem in the literature on cultural analyses of smartphones—the overlooked aspect of local diversity. The smartphone mediascape has recently been examined from the perspective of the convergence of different media (e.g., Goggin 2010; Sinckars and Vonderau 2012), yet only a few locally contextualized empirical studies address the locality of smartphone culture (e.g., Hjorth et al. 2012). In recent smartphone studies, the local dimension is addressed at best as a site in which a "global" technology is consumed and disseminated, while the production side of a local smartphone mediascape is largely ignored. Thus, despite the explosive influx of smart technologies into our daily lives, we know little about how the production side of a given local smart mediascape is articulated with the way in which local consumers appropriate such technologies. We are therefore required to empirically examine how smart media are produced, appropriated, and thus redefined in local contexts for a comprehensive understanding of what smartphones mean to youth culture. Through this multifaceted theoretical framework, I expect to shed light on the emerging debates on the smartphone's role in young Koreans' use of Kakao Talk.

Socioeconomic Perspectives of the Smartphone-scape

Apple's iPhones have led the global "smartphone fever" over the past few years, given that the brand has initiated and dominated the global smartphone market. However, it may be unobservant to confine the discourse on smartphones to iPhones (Sinckars and Vonderau 2012), as different smartphone handsets, technological infrastructure, regulations, and uses are increasingly emerging in various local contexts. Although the iPhone may have set the protocol of smartphones in the early period, they have been constantly localized over the past few years. In this respect, Korea's smartphone industry and culture offer an interesting example of the way in which smartphones are locally appropriated and defined by the industry and the consumers.

As discussed in previous chapters, the consumers' shift from feature phones to smartphones has been impressive in Korea. While there are several contributing factors, Koreans' phenomenal uptake of smartphones cannot be explained without considering the country's mobile media environments and enthusiastic ICT consumers, as well as favorable government policies, as already discussed. Above all, the timely introduction of high-speed data services and availability of cutting-edge smartphone hardware have contributed to boosting the domestic market. Korea's mobile telecommunication operators, such as LG UPlus and SK Telecom, one after another, have rushed to offer a faster data service known as LTE (Long-Term Evolution), followed by LTE-A, which has resulted in the bandwagon of the smartphone network (S. Yoon 2014b).[1] In addition to "wirelessly wiring" the country with enhanced data services, the domestic handset makers' research, development, and promotion have been factors in the local market expansion.

A socioeconomic perspective on the Korean smartphone industry's expansion shows that the product has been predominantly defined and marketed as mobile telephony or mobile Internet at best, while the technology's playful aspects and content have not been promoted well in the local market, which is controlled largely by major domestic corporations. The underdevelopment of creative smart media content has been implicated on the government's approach to the IT industry. In his address on the 67th anniversary of liberation, President Lee Myung-bak (2008–13) stated, "In the future smart society, creativity is the biggest growth engine, and the source of competitive edge," promoting his vision of a smart society with procorporate IT policies (Presidential Speeches [Taeťongnyŏng yŏnsŏl] 2012).

However, the rapid smartphone market expansion led by a handful of major Korean-based corporations has not necessarily been favorable to the rise of creative and participatory cultural economies sought by both IT-savvy young engineers and consumers. Large corporations have focused on telecommunication as a commodity, rather than exploring the potential of smartphones as a new phase of e-entertainment or gameplay. Despite the unfavorable media environment, a few venture corporations have been exceptionally successful in designing locally specific smartphone apps such as Kakao Talk, which have built a new user base. Kakao Talk is an outstanding example of how the perspective of smartphones as a form of telephony or Internet carrier, pervasive in the early period of the Korean smartphone market, has been challenged and redefined as a platform for not only "talking" but also "playing."

Meanwhile, it should be noted that the phenomenal growth of the smartphone market in Korea would not have been possible without consumers who have enthusiastically adopted and appreciated the playful aspect of the new technology. As discussed elsewhere (Jin 2010, 28) in the case of the broadband explosion, "Korea is full of 'early adopters' who are willing to buy newly released digital devices for consumer testing." It has been reported that Koreans are highly conscious of the need not to be left behind by others (Ward 2004), and the young have a great desire to conform to their peers' gadget-carrying norm. The smartphone-led mediascape has also made possible highly mobile and flexible communications for youth. Different media platforms are now reimagined and redefined on the touch screens of smartphones. In the early smartphone market's growth in Korea, smartphones were predominantly imagined as an upgraded mobile telephony, rather than a set of playful social media that may go far beyond the former's role.

Emergence of the Kakao Talkscape

As smartphones have become a major part of people's daily culture, as discussed in chapter 5, Korea has witnessed the rapid growth of the app economy—a range of economic activities encompassing the sale of apps, ad revenues, and digital goods on which apps are designed to operate (MacMillan 2009)—over the past few years. Kakao Talk, a virtual machine platform, has especially played a significant role in the rising smartphone-mediated youth culture in Korea. As the interviewees in my field study consistently expressed, Kakao Talk is one of the major reasons why Koreans have switched from second-generation (2G) phones to smartphones. According to my survey conducted in January 2015, when participants were asked about the major uses of their smartphones, Kakao Talk and other applications were the most frequently mentioned, at 70.9%, followed by news and information search (15.8%), mobile games (6.2%), Twitter and mobile Facebook access (3.3%), multimedia (3.2%), and others (0.6%) (fig. 12). As this data proves, Koreans use their smartphones mainly for Kakao Talk, both in terms of younger users and older users, because Kakao Talk is a must-have application for smartphone subscribers in Korea.

Kakao Talk enables users to access numerous free apps that can supplement Kakao Talk's messenger features. Kakao Talk allows person-to-person and group chats by simply entering users' phone numbers without any limit on the number of participants and without registering or logging in. In other words, Kakao Talk successfully addresses the small screen of a smartphone and the fact that it is often used on the move. It has a simpler interface than most consumers would expect, but its communication functions are sophisticated and convenient. Messages exchanged are shown on one screen in chat bubbles, and Kakao Talk was the first mobile messenger to accommodate group conversations.

Since its release in March 2010, the number of subscribers has rapidly increased (E. Choi 2013; Chŏng 2013), and Kakao Talk's growth forced Samsung to withdraw from the instant message service market. When its subscribers soared to 160 million in November 2014, Samsung's instant mobile messenger service ChatOn unofficially decided to terminate its service. ChatOn also had 100 million global subscribers thanks to Samsung Galaxy's popularity; however, people did not continue to use it because they simply switched to Kakao Talk in Korea, Line in Japan, and WhatsApp in the U.S. (*Dong-A Ilbo* 2014). This was rather shocking news for many telecommunications experts and smartphone users, in that Samsung's software, which is created and operated by Samsung, itself was not able to overcome the dom-

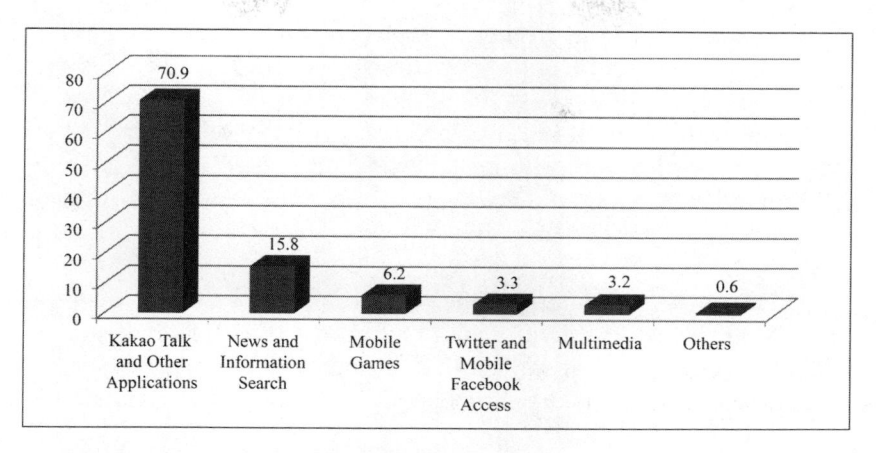

Fig. 12. Major Uses of the Smartphone (unit: %)

inant power of innovative software. This perhaps implies that the innovative ideas of venture capitals and individuals might be able to successfully compete with megatransnational corporations in the future as long as their innovative ideas can appeal to customers.

The large number implies that virtually every smartphone user is on Kakao Talk in Korea, and even its overseas user base is growing. As of March 2014, smartphone users accessed Kakao Talk on average 50 times per day and spent 46.42 minutes on Kakao Talk every day, which is the longest time among major apps, up from 43 minutes in 2012 (*Yonhap News* 2012; Pak and Kim 2014). Although Daum Cafe shows the highest access time at 66.08 minutes, it has a smaller number of users (table 8).[2] Young people are the most enthusiastic user group, as 56.9% of young Kakao Talk subscribers used the app for longer than two hours each day in a 2012 survey of Korean high school and university students (Yi Ch'ang-ho et al. 2012, 89).

Young Koreans especially tend to frequently use Kakao Talk's messenger feature, among several others features it offers. In my interviews, many respondents ascribed Kakao Talk's popularity to convenience and ease in its messaging services. A 28-year-old graduate student stated that Kakao Talk made it easy for him to make an appointment with friends, compared to contacting them by telephone or e-mail. A 23-year-old female university student said:

Kakao Talk is convenient, as it provides an online messenger function (thus I can send and receive files, and have a group chat) and a normal text function. It is more convenient [than other messenger apps], as it does not require any additional connection and login.

Of course, this does not mean that Korean smartphone subscribers use Kakao Talk exclusively, without accessing any other messenger apps. Young users tend to install, browse, play, and/or delete various apps and social games, as reflected in a 24-year-old man's account:

> My activities on the smartphone are all related to the Internet connection. My main activities are first, using Kakao Talk and Facebook, second, reading news articles through portal websites, such as Naver and Nate [Korean news portal sites], and third, checking and sending e-mails.

While numerous global and local smartphone apps compete with each other in appealing to Korean consumers, the local app Kakao Talk took the initiative in the domestic market.

Kakao Talk's remarkable growth has several causes. Above all, its timely release met emerging smartphone adopters' needs (Huh 2011). As a 24-year-old respondent noted, "The huge popularity of Kakao Talk is not because of technological quality, but just the first mover's advantage." Indeed, Kakao Talk as the first mover has succeeded in securing a large user base and thus dominating the local smartphone app market.

TABLE 8. Top Applications in Terms of Access and Time

Rank	Apps	Average Number of Accesses per Day	Average Access Time per Day (minutes)
1	Kakao Talk	50	46.42 minutes
2	Daum Cafe	23	66.08 minutes
3	Kakao Story	21	29.18
4	Facebook	19	41.15
5	Naver	16	34.25
6	Chrome	13	30.35
7	Naver Cafe	12	21.55
8	Africa TV	8	38.39
9	Anipang on Kakao	6	43.51
10	Cookie Run on Kakao	5	25.51
11	Naver Webtoon	5	20.43
12	YouTube	4	30.35

Source: Data from Pak and Kim 2014.

In this respect, Kakao Talk can be compared with global and domestic social messenger giants such as Facebook, Skype, NateOn, and Cyworld (practically defunct as of August 2016), which were originally designed for a relatively sedentary, personal computer (PC)-based setting and later had to adjust themselves to a new, smartphone-oriented media environment. The adjustment may be accompanied by "downsizing experiences" through which users have to reorient themselves to a small screen-based interface. Visual elements formerly displayed on an 11- to 20-inch screen are now minimized to a small touch screen at the expense of visual and interactional details. However, Kakao Talk is relatively free from such downsizing experiences, since it has an inherently smartphone-based interface.

Furthermore, Kakao Talk has quickly popularized its instant mobile messaging service by redefining the concept and practice of mobile messenger as both storytelling and gameplay, as reflected in the name and feature of one of its associated apps, Kakao Story. Kakao Story is a photo-sharing social network for Kakao Talk users released in 2012, and Kakao Story lets people share life as it happens with photos. Kakao Story has recently joined the SNS market, and it has grown into one of the most popular mobile-based SNSs in the domestic market in a short period of time, competing against Facebook and Twitter. Several respondents, many of whom were using Kakao Story along with Kakao Talk, tended to keep writing about their everyday lives on smartphones and wanted their micronetwork members to read and imagine what they do and where they are. Such media experiences of storytelling on smartphones are often presented to peers with photos taken by the device's camera and cute emoticons. Thus, one's everyday life becomes a drama or journey to be envisioned by Kakao Talk friends. In this respect, it can be argued that Kakao Talk revamps the strategy of earlier local social networking sites—Cyworld in particular (a Korean-based social network site, which in October 2015 became Cyhome)—in which personal messages are narrated and presented to one's micronetwork in virtual rooms (*pang*) where stories are told (Hjorth 2007b).

The emergence of the Kakao talkscape has been one of the most noteworthy media phenomena in Korea since the introduction of smartphones. As a locally specific social app, Kakao Talk has risen during the Korean IT industry's response to the global expansion of smartphone markets. As one of the first, natural-born, smartphone-based apps in the Korean market, Kakao Talk acted as a catalyst for redefining the smartphone as an articulation of telephony with gameplay and storytelling.

Kakao Talkscape as "Talking" Space

The smartphone and its locally contextualized mediascape, or the Kakao talkscape, have established an environment in which young Koreans manage and negotiate their limited temporal, spatial, material, and cultural resources. Smartphones are extensively appropriated for "microcoordination," or "nuanced instrumental coordination" through which young people organize their daily lives (Ling and Yttri 2002). In survey responses, it appeared that the smartphone was an essential means for its users to manage their daily schedules and social networks, as a 25-year-old female office worker at Samsung described:

> I use smartphone apps made by my company in order to check the commuting bus timetable, daily menus in my company cafeteria, and other updated information about the company. I also use mobile banking service and the T'imon [social shopping] app.

Among various smartphone apps that have particular functions for microcoordination, Kakao Talk seems to be the most extensively integrated into the rhythm of young people's urban lives. Young Koreans' use of Kakao Talk revolves around their management of sociality and urban space. On the one hand, they make efforts to maintain strong ties with their micronetworks and harmony with others, and on the other hand, they seek to organize their own time and space via smartphones, with a highly personalized use of the technology. Of course, as one male programmer in his late twenties explains, he has changed his phone from a feature phone to smartphone because it saves him a lot of money for communication, for example by using Kakao Talk, which enables free communication between people. However, since people download several apps, both paid and free apps, including mobile games, "saving money" is not a big deal for many Koreans. Instead, Kakao Talk provides a unique talking space, while also functioning as a playing space.

First, the rise of Kakao Talk and smartphones is part of the process through which young Koreans negotiate face-to-face communications. As Hjorth (2011b, 123) has noted, smartphones' relevance "is intrinsically linked to maintaining face-to-face social capital"; this tendency is evident among young Koreans who have vigorously adopted Kakao Talk's person-to-person and group chat features. In particular, group chatting was considered by most interviewees to be the app's most important feature. When asked how he would react to the absence of his smartphone for the next 24 hours, a 25-year-old man said, "I would be very anxious to hear from my friends, because I know that group chatting at Kakao Talk is going on without me."

Kakao Talk's group chat feature seems to substantially reduce young people's anxieties about being excluded or disconnected, since it gives them a sense of belonging all the time. A 25-year-old woman noted, "Kakao Talk is an essential tool for me to remain in group-chatting rooms with different group members." For most respondents, the management of existing micro peer networks was a key motivational factor in their adoption of smartphones (H. G. Lee 2011). Aware of the importance of belonging to affective micronetworks, known as *yŏnjul* or *chŏng*, Koreans tend to have enthusiastically adopted mobile technologies (K. Yoon 2003; H. G. Lee 2011). Moreover, harmonious communication is maintained in the Kakao talkscape. As observed in previous studies, East Asians tend to appropriate mobile communications in order to avoid conflicts that may occur in face-to-face interactions (e.g., Yuan 2012).

Second, the smartphone and Kakao Talk allow young people to customize the cityscape. As early mobile media researchers anticipated, the smartphone enhances personalized ways of appropriating communication technologies with password protection or filtering services to screen certain e-mails and calls (Kohiyama 2005, 70–71). However, the personalization of smartphones is not simply a reflection of individual users' desires, since it concerns a sociocultural context in which young people cope with limited spatial, temporal, material, and symbolic resources. Similar to Ito and Okabe's (2005) Japanese case study, the high-density urban population in Korea allows its youth few private places; they also have limited leisure time because of the competitive university entrance exam and the need for curriculum vitae development. Thus, the smartphone's various apps and convenient features of schedule management provide young Koreans with pragmatic and symbolic assistance (H. G. Lee 2011), which has also been called microcoordination (Ling and Yttri 2002).

Youthful appropriation of the smartphone as a personalized technology may correlate with the significant increase of "living alone" lifestyles among young adults in Korea. The rapid neoliberalization of higher education and the youth labor market has substantially undermined young Koreans' job prospects over the past several years. Consequently, larger numbers of young Koreans have postponed the transition from school to work and marriage. The new breed of young Koreans, called "self-cocooning young people" (*nahollojok*), seem to look for distanced, mediated communication rather than face-to-face engagement (H. G. Lee

2011). Several young adult respondents in the current study, who appeared to pursue a highly individualized lifestyle, described smartphones as a firmly personal technology rather than a device for socializing. According to them, the smartphone is an individualizing technology in comparison to other forms of ICTs, because it protects one's privacy, allows individual access to broadcast media content, and/or increases pseudocommunications. A 27-year-old graduate student in Seoul claimed that the smartphone's key function is to protect his privacy, as "it is not possible to figure out the subjects doing personal activities on smartphones." However, those who considered smartphones a means for effective privacy protection had limited awareness of the potential breaches of privacy on smartphones and Kakao Talk. Similar to many other social apps, Kakao Talk saves users' personal messages and stories on its main server, and thus it is possible for others to access and trace one's personal messages, as shown in the recent criminal investigations in which Kakao Talk messages have been adopted as evidence of crimes (Yi and Yi 2012).

The personalization of the smartphone is also observed in young people's "narrowing down" of the broadcast media. They use smartphones for individualized TV viewing by installing the apps of broadcasting stations, such as Pooq (by MBC), Gorilla (by SBS), and K (by KBS). These apps offer TV programs on smartphones, thus changing the concept of watching TV, as a 25-year-old male student stated:

> I don't like watching TV much, but I do view one or two programs regularly. It's a luxury for me to buy a TV only for myself or solely for watching a couple of programs. Instead, I watch my favorite TV programs on my smartphone, using these apps from broadcasting stations. I can watch all the programs free and in real time.

Traditionally, watching TV has been necessarily experienced by all family members, since TV has usually been placed in the living room, at the center of the home (Silverstone et al. 1992). Now, however, TV viewing is no longer a communal and sedentary experience but takes place anytime and anywhere, without any interaction with other viewers who physically co-exist.

Interestingly, a few respondents identified smartphones and Kakao Talk as tools of pseudocommunication, since they believed that the smartphone simply entices its users to take part in empty conversations and highly narcissistic self-presentations. A 26-year-old female respondent who recently had to change from a feature phone to a smartphone because of pressures from her peers and acquaintances noted that smartphones lead people to "meaningless," "time-consuming" chats:

> People update too often what they are currently doing via social media. People group-chat on Kakao Talk too much; they keep doing so everywhere, but there's no point in their talk, just full of individual voices. They do the same thing even when they meet people face to face.

For those few respondents who held somewhat skeptical views of Kakao Talk, the app interrupts the rhythm of one's daily life and may adversely affect one's face-to-face relationships.

Because of the rapid growth of instant mobile messenger culture, people's communication methods have significantly changed, primarily from e-mail-driven relatively formal communication to instant mobile messenger-driven informal chats. Several respondents said that they like Kakao Talk because it is very convenient for them to deal with lighter but more urgent tasks and issues, while using e-mail for any task or issue that should be put in writing. As they cannot log onto and check their e-mail account all the time sitting in front of a desktop PC, they depend on their smartphone for work at home, not for e-mails but for communication via instant messenger programs, including Kakao Talk. Consequently, Koreans have dramatically reduced their use of e-mail. According to the Korea Internet and Security Agency's Annual Survey on Internet Usage (2014, 9), out of the Internet users aged 6 and over, 59.3% are e-mail users who have used e-mail in the last one year in 2014, down from 85.7% in 2011. The percentage of e-mail users has been declining every year, from 85.7% in 2011 to 84.8% in 2012 to 60.2% in 2013, primarily because people prefer instant mobile messenger programs to e-mail for their informal communication (Korea Internet and Security Agency 2008; 2009; 2010; 2011; 2012; 2013; and 2014).

This trend is much different from the U.S., in which the number of e-mail users has continued to increase. According to eMarketer (2013), a consulting company in the U.S., among Internet users the percentage of e-mail users was 88.7% in 2012 and 89% in 2013. The prediction by eMarketer is 90.4% in 2017. By contrast, e-mail usage in Korea has been fundamentally influenced by the changing sociotechnological milieu in the smartphone era.

However, despite the few skeptical views on smartphone-mediated communication, overall, most respondents perceived the smartphone in

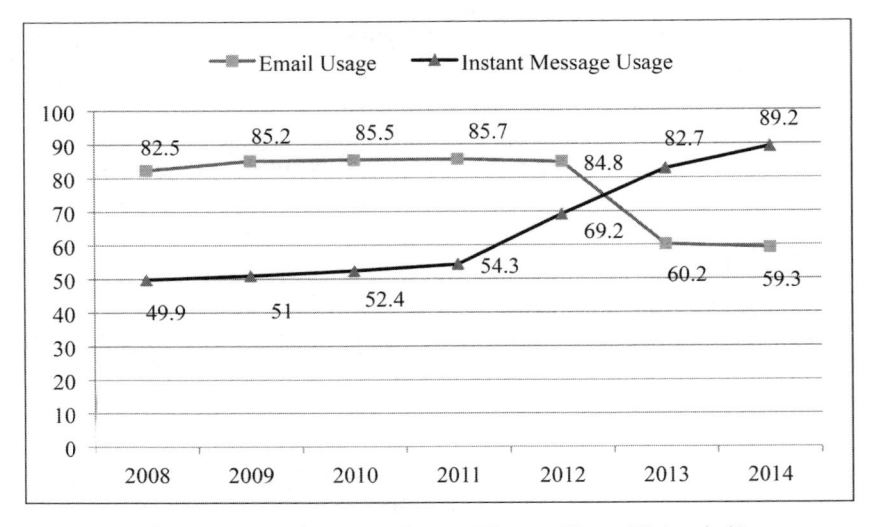

Fig. 13. Change in E-mail Usage vs. Instant Message Usage (Unit: %). (*Source:* Compiled from Korea Internet and Security Agency's [Han'guk Intŏnet Chinhŭngwŏn] Annual Survey on Internet Usage 2008, 2009, 2010, 2011, 2012, 2013, and 2014.)

its two seemingly opposing yet overlapping aspects—the socializing and the individualizing aspects. Given that technologies are appropriated by the individual yet domesticated by the social (Silverstone et al. 1992), the smartphone may be a *sociable* media technology that can also be *personalized* by individual users. This perspective implies that while the personalizing aspects of smartphones have been overemphasized in recent studies on new media (Bolin 2012), a closer observation of the localization of the technology shows a more complicated picture of users' engagement.

Kakao Talkscape as "Playing" Space

Kakao Talk has been a highly accessible gateway to numerous mobile social games such as Anipang, the most popular Korean puzzle game, played by an average of 10 million visitors each day at the end of 2012 (M. Cho 2012). Ordinary young people who would not play online games without smartphones and Kakao Talk have been introduced to casual and mobile gaming. In this respect, the Kakao talkscape challenges the stereotypical image of digitally oriented youth naturally equipped with spectacular gaming skills and play. It should also be noted that studies on mobile technologies have tended to overstate the affinity between youth and technology, and the subcultural nature of youth. Thus, essentialized images of mythic "mobile youth," who are inherently technology savvy and politically liberal, have been reproduced (e.g., Tapscott 1998). In order to move beyond the stereotype, it may be cogent to be aware that young people's media practices are not necessarily "spectacular," but increasingly embedded in the ordinary routine of everyday life. The Kakao talkscape is especially interesting in that it reveals how "ordinary" young people engage in an emerging mobile media environment. Kakao Talk has attracted an increasing number of ordinary ICT users to mobile gaming, while recreating mobile messenger and social networking services through its relatively playful, visual, and open interface.

The vibrant mobile gaming culture observed in the Kakaoscape may be a product of a conjunction of Koreans' community-driven play patterns and the country's high rate of urbanization. Community-driven gameplay, identified as unique to Korean gamers (Jin 2010), is also observed among many Kakao Talk users. Interestingly, mobile gaming, which provides a highly individualized and personalized gaming environment, especially in comparison to the sedentary, PC bang-based online gaming culture in Korea, has not been fully individualized in Korea.

Urban space is redefined by young people's cultural practices via smartphones, in which the exchange of affection and data is blended with gameplay. In other words, in the renegotiated urban space, mobile gaming is immersed in everyday communications, while text messaging becomes a form of gameplay. The phenomenal rise of Kakao Talk as a playful messenger and gateway to several social games has led young people to various "smart" texting practices on touch screens, which implies a "social struggle over appropriate usage" of a new technological form (Ito and Okabe 2005, 132). While swinging between conformity to the dominant communication norms and the pursuit of individuality, young people relax power relations involved in face-to-face communication. Korean youth may especially benefit from digitally mediated written communication that may lessen the strict age and power hierarchy embedded in the Korean language (Herring 2003). By negotiating the traditional linguascape, young people may exercise "semiotic democracy," in which everyday communication can potentially be transformed into polysemic and open texts (Fiske 1987).

Young people's engagement in texting practices on smartphones can be distinguished from text messaging on 2G phones, since the former involves sensual and playful experiences. In comparison to the experience of pressing keypads in the text-based 2G phones, stylish and playful handsets seem to provide young users with intimate and interactive feelings and fun. For example, a 25-year-old female respondent in Daejeon pointed out the sensual aspect of smartphones in comparison to other forms of ICTs:

> I played Angry Birds on an iPad and enjoyed making touch inputs. Smartphone games allow sensuous pleasures from touch inputs. I tried playing Kart Rider on both smartphone and PC, and it was more fun to play it on the smartphone than on the PC.

It is not only the handset but also the software app and the interface that differentiate smart media experiences from those of other media technologies. A 25-year-old man was aware of the variations in interface among ICTs:

> My feelings when playing games are totally different [depending on where I play]. Playing mobile games on a smartphone feels more like playing video games. Smartphone games also use touch inputs, so it gives me different feelings compared to when I use keypad buttons. . . . The touch input makes the biggest difference, I guess.

As media analysts and the Kakao Talk developers have agreed, the app's remarkable success is indebted partly to its open-concept interface (Huh 2011), which allows smartphone users to access and exchange messages and data easily with anyone on their "friend lists." Kakao Talk also allows for the convenient and free exchange of a wider range of emoticons, signs, images, and video files through its visualized interface. In particular, Kakao Story, Kakao Talk's storytelling app, has three cute emoticons to express and share users' feelings constantly with their friends on smartphones. Many respondents were regularly narrating their stories and reading their friends' anecdotes on Kakao Talk and Kakao Story. A 25-year-old man noted, "Without my extra effort to directly ask my friends, messenger programs such as Kakao Talk on smartphones naturally let me grasp what my friends did today by showing their uploaded photos or comments."

The Kakao talkscape shows how gameplay, texting, and storytelling converge and are practiced as experiences of playful communication via smartphones. Kakao Talk's significance lies in its affordance through which young users can use both gaming and messaging features, thus blurring the boundary between mobile telephony and mobile gaming. With the increasing sizes of touch screens, as Apple iPhone 6 and Plus as well as Samsung Galaxy 5 and Note prove, smartphone makers have continued to develop large screen-embedded devices because visuality has become one of the most important features for people when choosing their next smartphone.

Conclusion

This chapter has contributed to a nuanced understanding of the smartphone as a cultural and technological form by exploring both the technological convergence and cultural divergence of smart media, which is a pressing yet overlooked issue in the mobile media literature (Watkins et al. 2012). From a socioeconomic perspective, the Kakao talkscape is in part an unexpected outcome of the phenomenal growth in Korea's domestic smartphone industry. The Kakao talkscape shows how the smartphone is localized through an articulation of socioeconomic structure and agency. Major national IT corporations have set the highly concentrated infrastructure for mobile telecommunications in Korea, especially in response to the rise of iPhones.

However, despite the remarkable expansion of the local smart media market led by Korean-based IT corporations, smartphones were imagined largely as telephony until young ICT users engaged in creative local apps such as Kakao Talk. This app paved the way for the technology to be locally recontextualized, as its users have begun to imagine smartphones as a tool for telephony, stories, and gameplay. In particular, young Koreans' engagement in the Kakao talkscape has rejuvenated text messaging, which used to be a popular and youthful means of communication, as an "appropriate" urban communication custom and simultaneously opened a new door for social gaming as mobile, casual, and ordinary cultural practices. Young people's appropriation of Kakao Talk shows that communication is incessantly textually and visually mediated and further rendered playful, while mobile telephony and gaming are blended with each other. Face-to-face communication is transformed into playful practices in the Kakao talkscape. Kakao "talk," which is not literally a form of verbal exchange but a digitized, visualized, and mediated mode of communication, makes talk more sociable and personal on the one hand, and more playful on the other hand. Kakao Talk's potential to plasticize the hierarchical rules of the Korean language suggests that the rise of smartphones in the country provides resources through which ordinary young people negotiate dominant social, material, and symbolic power. In this regard, young people's engagement in the Kakao talkscape can be interpreted as a "subcultural bricolage" through which they adopt and rearrange available resources (Clarke 1976).

This chapter also suggested that despite its global popularity, smart media technology is being substantially localized. The Kakao talkscape vividly shows how different media technologies converge in the platform of smartphones, yet how divergently such a phenomenon may occur in varying local contexts. As an open gateway to various other online services and apps, Kakao Talk has responded to, and stimulated further, ongoing media convergence. Kakao Talk's role as a go-to messaging service for any smartphone subscriber has significantly appealed to many local users looking for a new mode of social media and for a gaming platform as they switch from 2G phones. The phenomenal local uptake of Kakao Talk resonates with Jenkins's (2006) assertion that media users' migratory behavior in pursuing new trends has become a primary reason for media convergence. The Kakao talkscape addresses a divergent reality of convergence, in which a new technology is negotiated through community-based local modes of communication and urban space, embodying the country's rapid modernization.

9

Beyond Smartland Korea

The smartphone has become one of the most significant new technologies of social mediation in the early 21st century. There have been several remarkable new media technologies in our modern society, including television, telephone, and the Internet; however, the smartphone is one of the most significant, if not *the* most influential, thanks to its functionality based on the convergence of technologies and daily activities. Smartphone users in 2017 may not imagine what the first smartphone looked like, yet the history of the smartphone is not that extensive, having started only about 20 years ago. From the bulky IBM Simon that weighed 18 ounces and cost around $899 in 1994, to the sleek, feather-light, and affordable smartphones of today, smartphones have undergone an inspiring evolution over the past two decades (Harjani 2014). When the London Science Museum celebrated the 20th anniversary of the birth of IBM Simon, the first smartphone, in 2014, it stated:

> You probably didn't use it, and may not have even heard of it, but the now 20-year-old IBM Simon was the world's first smartphone, on which you could write, draw, manage your contacts and calendar, and send faxes—it even had apps. It was the first device to merge the mobile phone and the personal computer, and so to celebrate Simon's 20th birthday, the London Science Museum is putting it on display in its new Information Age exhibition alongside the inaugural BBC radio broadcast in 1922, the first digital TV and a further 800 breakthrough technologies. (*Independent* 2014)

Today almost everyone uses, sees, or at least has heard about the smartphone, and the dramatic improvement of smartphones within such a short period of time has substantially influenced people's daily lives, including the economy and culture. Only a few years ago, people used mobile feature phones for person-to-person communication; but "the smartphone has reinvented the communications landscape. By individualizing communication we can be in contact with our nearest social sphere whenever and wherever we (and they) may be" (Ling 2012, 157–158). Smartphones are now like extensions that people cannot separate from their bodies, as people stay near their smartphones almost 24 hours per day. The smartphone has not only highlighted the relationship between space and communication but also developed "social relations across time and space" (Jensen 2013, 26). In other words, the smartphone with a touch screen is not only a personal communication tool but also a social network converging necessary functions, from e-mail to digital camera to GPS, and smartphones and apps play a key role in developing our digital economy and culture.

As the diffusion of smartphones has occurred globally since 2007, when Apple started to sell its first iPhone in the global market, smartphones have profoundly rearranged the social milieu of mediated communication, and smartphones have moved from being useful to being essential, and are now taken for granted. Smartphones have changed the ways we coordinate everyday life, and they have shifted business and commerce: the app economy is growing, mobile gaming is getting more popular, and Kakao Talk is a must-have app in Korea. All of these changes have been made possible by the dissemination of smartphones. As with many other cutting-edge technologies, including broadband and online gaming, Korea has become one of the centers for smartphone technologies and culture. All of these phenomena emerge from a technology that was generally unknown only a few years ago (Ling 2012).

This book has examined the recent emergence of the smartphone and related smartphone culture in Korea by addressing several key issues, including distinctive information technology policies, protective measures in the midst of neoliberal telecommunications policies, severe competition among IT corporations, and enthusiastic IT consumers. I looked into a hitherto neglected focus of inquiry, a localized landscape emerging with smartphone apps, including mobile games and Kakao Talk. I focused on not only the celebratory achievement of local smartphones, but also the significance of social environments in the development of smartphones. I situated the emergence of smartphones within the growth of mobile technologies and overall telecommunications industries embedded in Korea's information and communication technologies.

Because of the smartphone's significance as one of the recent symbols of globalization, I also attempted to determine the role of local-based technologies and cultures in the era of neoliberal globalization. Several global actors, including both Western countries and Western-based transnational corporations, including Apple, played a significant role, particularly in the first stage of the smartphone. Therefore, I critically investigated the power struggles between global and local forces in order to determine whether the emergence of local smartphones has changed the contours of globalization theory.

In this book, three major areas of inquiry, including innovation and the evolution of the smartphone; digital economy through the lens of political economy; and people's culture, including youth culture, embedded in the Korean smartphone context converged in order to comprehensively address the localized smartphone landscape. It is essential to enter into a discussion of emerging smartland Korea by utilizing diverse perspectives, given that technology is no longer an isolated area of study. I emphasized that the growth of smartphones must be defined based on the sociocultural specificity of Korean smartphone usage. These holistic approaches interweave in the emergence of the smartland Korea phenomenon, whose major characteristics can be summarized as follows.

I found several dimensions to the growth of the smartphone, where Korea has moved from the Apple iPhone shock to a smartphone wonderland within a short period of time. To begin with, the Korean government has developed a unique mobile telecommunications policy. In the midst of neoliberal globalization, emphasizing a small-government regime, the Korean government has actualized state-led developmentalism. As "technological changes can be both helped and hindered by economic and regulatory conditions" (Havens and Lotz 2012, 49), one of the most significant factors deciding the contours of the smartphone industry as part of telecommunications has been favorable government policies. External neoliberal pressures from the U.S. government and transnational corporations and internal business interests from telecommunications corporations and service providers have pushed the Korean government to take significant deregulation and liberalization measures; however, as shown in the case of the delay of the launch of iPhone in Korea, the government has continued its primary role as a facilitator and regulator. Korea's smartphone technologies and relevant policy issues have not only reflected neoliberal telecommunications policies, but also Korea's state developmentalism as part of the continuous growth of the local telecommunications system. While neoliberal globalization gains support in many countries, it does not eliminate the crucial role of the nation-state in the realm of Korea's smartphone sector.

Second, I developed new perspectives in the existing body of knowledge on the issue of globalization by discussing its pertinence to smartphones. By addressing the theoretical question of globalization in the era of smartphones, from Korea's reception of the iPhone in 2009 to the recent growth of Samsung's Galaxy in the global market, I have shown that globalization based on mobility and connectivity has signified a power shift of capital, and forced local states to affiliate with or integrate into a part of the new world system. As Kwang Suk Lee (2008) argues, the survival of small countries depends largely on their close links to the global electronic conduits of capital, and Korea's smartphone growth can be situated within the universal structure of the electronic empire, which we revealed as desperate, striving to enlist the local as an active part of the new global network.

Third, I found that the emergence of the app economy has depended not only on infrastructure but also on software, and that the evolution of smartphones has consequently influenced the software sector in the Korean market. After analyzing the rapid growth of smartphones and apps in the socioeconomic milieu specific to the country, including several key aspects of smartphones and apps services in Korea, as well as their implications, I concluded that Korea has become the world's best laboratory for smartphones and apps—a place to look to for answers on how the app economy may evolve. In other words, this book has recognized technology as a socioeconomic product that has historically been constituted by certain forms of knowledge and social practice.

Fourth, I provided a new approach to the digital divide by discussing its pertinence to smartphone technologies. Based on an empirical survey, I developed a new perspective, the dual divide, in addition to the traditional divide, in access to and use of new technologies, in comprehending the smartphone divide. I proposed that the digital divide, particularly in the realm of the smartphone, can be understood primarily through social inclusion, because the smartphone divide cannot be analyzed only through our conventional dichotomy between the device haves and the device have-nots. Scholarship studying the smartphone divide needs to focus on the ways in which smartphone users benefit from the position and use of the new device.

Finally, I concluded that smartphone culture has taken shape within the context of Korea's particular mobile culture. By exploring some of the sociocultural factors contributing to the growth of smartphones in Korea, I demonstrated that the sociocultural milieu surrounding the rapid growth of smartphones and apps, focusing on two major areas—mobile gaming and Kakao Talk—has contributed to the growth of mobile gaming and youth culture pertinent to the smartphone. On the one hand, the emergence of smartphone use has shaped the development of Korea's mobile games, compared with what one might find in other use cultures and national contexts. Smartphone users have perceived their own changing conditions amid the process of rapid diffusion and growth of mobile games. On the other hand, I found that a localized media environment, arising with the smartphone and its apps, with particular reference to young Koreans' engagement in a local app platform, Kakao Talk, has created the "Kakao talkscape" as a form of mediascape embedded in and constructed by specific sociocultural circumstances. Smartphone users have engaged with the rapid socioeconomic diffusion of smartphones and Kakao Talk.

Accordingly, smartland Korea provides a unique test bed for global mobile experts and policymakers, primarily because of people's increasing loyalty to their smartphones. Koreans' adoption of new technologies is faster, their use of the smartphone is much broader, and their reliance on the apps is more embedded than those of people in other countries. Smartland Korea, however, is not analogous to an idealized Disneyland consisting only of full of excitement and fun, because there are several social issues in smartland Korea, including the smartphone divide. Although Korea has rapidly shifted its own digital economy toward the app economy, U.S.-based operating systems are the primary beneficiaries as the dominant players. While enjoying the smartphone, we should not ignore the political and economic systems that shape the engagement between smartphones and ourselves (White 2014). It is critical to understand the next move in smartland, because people's daily activities will continue to be influenced by the changing socioeconomic and cultural milieu as the smartphone evolves and is embedded.

Smartland Korea will be confronting new challenges, in two different dimensions. On the one hand, what people can do with the smartphone will evolve in the future, and there are already several new developments that will be actualized in people's daily activities, as well as in our economy and culture. As CNN reported (2014),

> Sales of mobile devices are still growing but quickly leveling off. The technology industry is rushing to figure out what the next big game

changer will be, and major companies are betting big on small, wearable devices, referring to a broad category that can include fitness trackers, smart glasses, smartwatches, clothing with embedded sensors that gather data while zig zagging their way through your lower intestine. However, most wearables are not meant to replace smartphones. Instead, they work as satellite devices that amass useful data or relay notifications from a primary mobile device. If they have screens, they can display simplified versions of mobile apps.

Furthermore, mobile devices will have the ability to understand who and where their user is, and these devices could track identity, location, and context, and feature flexible screens and enhanced battery life.

Several corporations have already moved ahead. For example, Samsung announced its new project, known as Smart Home through "Internet of Things" (IoT) services. This project will connect refrigerators, washing machines, televisions, camera phones, watches, and more, and people will be able to control their home devices from a smartphone and/or a wearable device (Siegal 2014). In January 2015, the Korean government announced that it will provide loans of $92.3 billion for the development of the IoT and other software projects to foster new growth engines. The government said that the state-financing scheme will benefit fifth-generation mobile networks, biomedicine, solar and fuel cells, bioenergy, and nano-semiconductor and sensor technologies (S. Yoon 2015). Since the smart home market will grow 19% on average per year between 2014 and 2019 (Yi Chin-myŏng 2014), it should become a main part of the app economy; therefore, transnational corporations once again have no choice but to invest resources to become leaders. This means that the smartphone will be the most significant cutting-edge digital platform over the next several years, and it is certain that there will be several new features added to the current form of smartphones.

On the other hand, Smartland Korea also asks policymakers and researchers together to develop new forms of social inclusion because there will be new forms of the smartphone divide due to the increasing gap between heavy users, who enjoy these new features, and general users who use basic features, resulting in the increasing gap in their economic and cultural lives. In particular, middle-class incomers have continuously fallen behind those higher up on the income spectrum. That is a serious challenge that Korean society has to take on, because these lower-income people may not be able to enjoy new smartphones and apps, resulting in increasing inequalities. Since the majority of app developers also target Google Play and Apple Store in order to enhance their potential success, the dominant role of the U.S.-based operating system providers and software start-ups will continue unless emerging businesses, such as corporations in China and Korea, are able to develop both comparable operating systems and apps that will be widely used globally.

Overall, on the topics of the smartphone and therefore media-ICT issues, it is critical to develop new perspectives explaining local contexts, which should involve a more nuanced and complex understanding of the growth of the smartphone and the app economy, as well as smartphone culture. Until a few years ago, academic discussions of mobile communication were marked by the primacy of North American and European models, and the particularities of other regions and countries tended to be elided (McLelland 2007; Tai and Zeng 2011). However, with the rapid growth of smartphone use and relevant apps, this formerly neglected region cannot be ignored anymore; the smartland Korea discussed in this book proves the significance of the local landscape.

It is also crucial to comprehend how interests and cultures clash and hybridize on the micro level and how local people develop their playfulness. As illustrated by this book, it is in a holistic approach that we find analytic strength. In other words, "A firm grasp of the big picture may enable cultural studies to foreground the issue of structural power in their inquiries of local audiences or consumers, while a cultural turn may allow political economic studies to address the criticism that they have neglected human creativity and lack empirical evidence on the micro, everyday level" (Shi 2011, 151).

One last thought is that we have to think about how key features of the global mobile system have been reorganized and transnationalized since the early 1990s, that is, how the transformation of the global mobile system can be understood within the larger context of global political-economic shifts and accompanying technological culture. I therefore expect that this study of Korea's experience of sociocultural and political-economic change within the smartphone system will shed light on more general trends in the global mobile system as it inevitably shifts in the near future.

Notes

CHAPTER 1

1. Kakao Talk allows person-to-person and group chats, by entering users' phone numbers, without any limit on the number of participants and without registering or login.

2. Line was released in 2011, although Line was invented by NHN's (Korea's largest web portal) Japanese subsidiary company.

3. Ito et al. (2005, overview), for example, argued that Japan's enthusiastic engagement with mobile technology has become part of its trend-setting popular culture. They covered the transformation of *keitai* from business tool to personal device for communication and play, and in this book I document the emergence, incorporation, and domestication of mobile communications in a wide range of social practices and institutions.

4. Since I emphasize the critical nexus of global and local forces in the smartphone sector, a few chapters use one or two major theoretical frameworks, in particular globalization theory.

CHAPTER 2

1. This chapter examines a multiplicity of source materials, with the majority coming primarily from printed sources. Documents such as annual reports and informational publications provide important information. Corporate data, including corporate archives, also have merit in that they provide detailed data on industry. However, IBM, which developed Simon, has no record of Simon in its archives. In these 1990–95 documents, there is no word on Simon. Since BellSouth was merged with AT&T in 2006, AT&T Archives is the only archive keeping information on Simon, but information is very limited. Meanwhile, trade journals are also fundamental sources for analyzing mobile and computer industries involved in the development of smartphone technologies. Trade journals are significant sources in this case because they provide crucial materials that cannot be accessed elsewhere. Although some trade publications present unsubstantiated and superficial coverage of corporate activities, reflecting press releases provided by industry public relations, trade publications give detailed descriptions of the historical activities of the media industry.

2. As Brey (1997, 4) documented well, "The term social constructivism is sometimes used in a narrow sense, to refer to the influential Social Construction of Technology (SCOT) approach that was outlined originally in Pinch and Bijker (1984) and Bijker (1987), and a number of related approaches, such as those of Collins (1985) and Woolgar (1991). In a broader sense, the term also includes what are called social shaping approaches (MacKenzie and Wajcman, 1999; MacKenzie, 1990) and the actor network approach of Bruno Latour, Michel Callon, and John Law, and their followers." In this chapter, I primarily employ the broader sense of the SCOT.

3. For the Genesis, the basic unit carried a suggested retail price of $350; cartridges cost $29.95 to $49.94 apiece and the module about $250, while the suggested retail price of the Touch-A-Matic 1600 was $149.95.

4. Motorola's 1978 DynaTac was the first cellular phone, invented by Martin Cooper and his team (FCC 2005; Zheng and Ni 2006, 32).

5. Many successful diffusions of innovative technologies have required several stages, as the diffusion-of-innovation theory, mainly developed by Rodgers (1983, 5), explains. The diffusion of innovation refers to "the process by which an innovation is communicated through certain channels over time among the members of a social system." As Rogers points out, diffusion is "a special type of communication, in that the messages are concerned with new ideas, and communication is a process in which participants create and share information with one another in order to reach a mutual understanding." The process can be explained through four different stages: (1) an innovation; (2) communication through certain channels; (3) over time; and (4) among the members of a social system (Rogers 1983, 10–11). However, the IBM Simon could be considered a successful failure: it was successful because it was a breakthrough technology, but it failed because it did not receive a meaning adoption, which is a decision of "full use of an innovation as the best course of action available" (Rogers 2003, 177). Therefore, while considering Roger's diffusion-of-innovation theory in understanding the early stage of innovation, it is critical to utilize the social constructivist approach in understanding the major players in the process of the invention and early growth in the context of socioeconomic milieu of that time.

6. Back then, manufacturers were struggling to make the communication devices market take off. The most common approach was to bring communication to what was then state of the art in portable technology: PDAs. Many products came out at that time, including the Marco and the Envoy, respectively running the Newton and the MagicCap OS: The Sony Magic Link. Most of those products were discontinued after no more than two years. Most of them featured large displays, and a notebook-oriented usability. They were basically PDAs with integrated (radio) modems.

7. During the first several years after Bell invented his telephone, the major users of this telephone were corporate. For example, in 1876, the U.S. had 3,000 telephones, and they were all owned by the Bell System. When the U.S. had 266,000 telephones in 1893, the Bell System owned almost all of them (U.S. Department of Commerce 1975).

8. The Stockholm Smartphone team was a group of former Sony Ericsson employees who had worked at the Stockholm/Kista site with smartphone development.

9. The AT&T Eo Personal Communicator 440, formerly known as the Eo Personal Communicator—shorter, but still not as catchy as "Newton" or "Zoomer"—was a pen-based portable computer that found its greatest success among workers and executives who needed to fill out forms, keep databases, send and receive faxes, check electronic

mail, and make telephone calls when there were no phone lines around (Lewis 1993). Its compelling rationale, for many, was the combination of wireless communications and computing. The AT&T Eo computer did all the things that the Newton and Zoomer promised to do, including sending fax and e-mail messages. Microsoft also proposed its own pen-software standard based on Windows (Lewis 1993). Eo Inc., which was acquired by AT&T, was working on a smaller, cheaper model designed as a smart cellular phone, more in the mode of Simon. AT&T bought McCaw Cellular Communications, the biggest cellular operator, and had a broad line of cellular phones that could mesh with the new Eo. But Eo had much going against it. AT&T was holding back on investing (Johnson 1994). Almost at the same time, Apple in 1993 spent more than $3 million on ads to introduce the Apple Newton MessagePad. Apple was working on improved models, but it did not yet offer a smartphone.

10. The wireless Internet flopped spectacularly in every part of the world except Japan. WAP, the wireless application protocol that was supposed to put cell phone users on the Internet in the U.S. and Europe, is memorable mainly for having inspired the slogan "WAP is crap." Yet i-mode, introduced with minimal expectations in February 1999, attracted more than 25 million subscribers—one-fifth of Japan's population. The Internet is never mentioned in the ads the public saw; the *i* in i-mode stands for information, and the logo—a large, stylized *i*—plays off the *i* that marks the information booths in subways and airports. Japan's infatuation with English-language product names even extends to DoCoMo itself: Ads proclaimed it to be an acronym for "Do communications over the mobile network," but *dokomo* is also a word in Japanese. It means "everywhere" (Rose 2001).

11. Inside the clamshell-style case was a chiclet QWERTY keyboard, complete with function keys for the major features and a series of programmable buttons by the screen.

12. The Communicator was a mobile powerhouse, with 8MB of memory and a 33MHz processor. This combination ran Nokia's own GEOS operating system (a predecessor to the Symbian OS used on later models), combined with a suite of business programs that could read and edit Microsoft Office files from a desktop PC. The screen was a black-and-white LCD, with a then-high resolution of 640 × 200 pixels. This long, thin screen meant that it could offer a first: a graphical web browser on a mobile device

CHAPTER 3

1. As of mid-1996, Korea restricted foreign direct investment (FDI) in 120 categories, including production and trading in agricultural goods, publishing, utilities (power generation and water), transport, telecommunications, banking, insurance, broadcasting and legal services. The Korean government, however, enacted the Foreign Investment Promotion Act in mid-1998 to entice FDI (U.S. Department of Commerce 1999). In June 2000, only four sectors, including radio broadcasting and television broadcasting, remained closed to FDI, and 17 sectors, including cable broadcasting (allowing up to 33%) and wire telegraph and telephone, and news agency activities (allowing up to 25%) were partially opened to FDI. (Ministry of Finance and Economy of Korea 2003; Jin 2011).

2. The importance of mobile communication was first accounted for in the 1980s, as worldwide demand for wireless communication devices surged, and new communica-

tion technologies were introduced. To respond to the changes occurring within the communications industry, the Korean government established Korea Mobile Communications Services (predecessor of SK Telecom) as a subsidiary of Korea Telecom (KT) in 1984. In its initial stages, Korea Mobile Communications Services provided car phone services in urban areas based on analogue cellular service technologies (Korea Trade Investment Promotion Agency n.d.).

3. Although Korea Telecom was a public enterprise, the MIC used its licensing power to control KT and to protect CDMA technology. "For the government, CDMA digital technology standards could provide it with a device with which to control the mobile telecommunication industry and prevent the penetration of foreign companies into the domestic market" (Jho 2007, 130).

4. The nine new growth engines included several key areas the government planned to focus on, including mobile communication, digital TV, home networks, and digital contents.

5. Cheol Gi Bae (2014) interviewed a KCC officer and revealed that the government had delayed the abolition of WIPI in order to protect domestic handset manufacturers. The officer said, "If we had abolished WIPI in 2007 and allowed iPhone to enter the market, could Samsung have succeeded in making Galaxy? We asked Samsung how long would it take to develop a smartphone like iPhone and Samsung answered it would take 9 months. The government's role was crucial in enabling Samsung to prepare the era of the smartphone. That's why a control tower is needed in Korea."

6. Until 2009, Samsung and LG together accounted for more than 85% of the domestic mobile market with their home field advantage. Their combined domestic market share for a while dropped to 61% at the end of November 2010. Samsung Electronics, seeing Apple's forecast-beating popularity on its home turf, tightened its ties with Apple's rising rival, Google, making a push into the smartphone segment with its Android-based Galaxy S smartphone. LG Electronics, the world's number three mobile phone maker after Nokia and Samsung, replaced its chief executive and head of "its mobile business team as it reeled from a record loss from its phone business" (*Yonhap News* 2010).

CHAPTER 4

1. When Apple released its first iPhone in other countries, the company indeed selected only one partner per country. Apple selected T-Mobile, the leading network operator in Germany, as the exclusive carrier of the iPhone when it made its debut in Germany in 2007. Likewise, Apple chose O2, the leading wireless carrier in the UK as the exclusive UK carrier (Apple 2007a; 2007b).

2. In terms of the history of the Internet, there are three major different starting points in Korea. The first is when KIET and Seoul National University were connected through SDN in 1982; the second is when Korea registered the national domain name .kr in 1986; and the third and final is when the global connection between Korea and Hawaii in the U.S. started in 1990. While it depends on the purpose of the research and study, this book considers the registration of the domain name in 1986 as the starting point of Korea's Internet history (see Yi Hyŏng-gyŏng 2014).

3. Outsourcing is the process whereby certain elements of production are completed

somewhere overseas in order to take advantage of cheaper labor conditions and government subsidies. Typically, only certain elements of production get outsourced, particularly those elements that require the greatest degree of labor and the slightest degree of creativity (Havens and Lotz 2012, 238–240).

CHAPTER 5

1. The term originally came into use in Korea around 2003 to describe what were previously known simply as Internet comics. The domestic market had rapidly increased to $151 million in 2013, and reading webtoons has been one of the major activities of many people (S. Kim 2014).

2. "Both apps and mobile websites are accessed on a handheld device, such as smartphones (e.g., iPhone, Android, and Blackberry) and tablets. A mobile website is similar to any other website in that it consists of browser-based HTML pages that are linked together and accessed over the Internet (for mobile typically Wi-Fi or 3G or 4G networks). The obvious characteristic that distinguishes a mobile website from a standard website is the fact that it is designed for the smaller handheld display and touch-screen interface. Like any website, mobile websites can display text content, data, images, and video. They can also access mobile-specific features such as click-to-call (to dial a phone number) or location-based mapping. Apps are actual applications that are downloaded and installed on your mobile device, rather than being rendered within a browser. Users visit device-specific portals such as Apple's App Store, Android Market, or Blackberry App World in order to find and download apps for a given operating system." The app may pull content and data from the Internet, in a fashion similar to a website, or "it may download the content so that it can be accessed without an Internet connection" (Summerfield 2014).

3. Kakao was listed on the Korean Stock Exchange on October 1, 2014, and put Beom-soo Kim (Kim Pŏm-su), founder of Kakao Corp., among the country's ten richest people (Mac 2014).

4. Kakao Talk was temporarily struggling, as of October 2014, thanks to a potential crackdown by the Korean government. Prosecutors announced the crackdown at the end of September 2014 after President Park Geun-hye complained about insults directed at her and said false rumors divided the society. "That rattled users of Kakao Talk and prompted a surge of interest in a previously little-known German competitor, Telegram. Rankey.com, a research firm, said an estimated 610,000 Korean smartphone users visited Telegram on October 1, 2014, a 40-fold increase over September 14, before the crackdown was announced." Telegram was the most downloaded free app in Apple's App Store in Korea on October 3, 2014. On Google's store, Telegram was the number two most downloaded free communications app, behind only Kakao Talk. Korean users left reviews on Telegram saying they left Kakao Talk to seek "asylum." "The uproar threatened to slow adoption of social media or send Korean users to foreign services, undercutting government ambitions to build a high-tech creative economy" (Associated Press 2014). This incident proves the vulnerability of the instant mobile messenger app market, because users are willing to change their major messenger app unless it is secure.

5. In the In-App Purchase model, Apple receives 30% of the purchase price and the

developer receives 70%. As Apple's guidelines (2014) explain (to app developers), "In-App Purchase gives you the flexibility to support a variety of business models in your iOS and OS X apps. With In-App Purchase, you can offer your customers additional digital content, functionality, that services and even subscriptions within your paid or free app. For example, In-App Purchase will allow you to sell: digital books or photos, additional game levels, access to a turn-by-turn map service, and subscriptions to digital magazines or newsletters. In-App Purchase is implemented in your app via the Store Kit framework."

6. Breaking down the operating system market for smartphones, again, Android is the largest in the world, and Korea is no exception. Although Korea is one of the major countries providing smartphones in the global market, no single smartphone maker has its own operating system. In Korea, previously, Android comprised only 6% of the market in 2010; however, it soared to as much as 93% in December 2013. In 2013, Korea was the only country with a figure above the 90% mark, because Samsung, LG, and Pantech focused on Android-powered smartphones (*Seoul Economic Daily* 2012; *Yonhap News* 2014b).

CHAPTER 6

1. The three categories in the ICT development index are as follows: (1) *Access subindex:* This captures ICT readiness and includes five infrastructure and access indicators (fixed-telephone subscriptions, mobile cellular telephone subscriptions, international Internet bandwidth per Internet user, percentage of households with a computer, and percentage of households with Internet access); (2) *Use subindex:* This captures ICT intensity and includes three ICT intensity and usage indicators (individuals using the Internet, fixed [wired] broadband subscriptions, and wireless broadband subscriptions); and (3) *Skills subindex:* This captures ICT capability or skills as indispensable input indicators. In the absence of data on ICT skills, it includes three proxy indicators (adult literacy, gross secondary enrollment, and gross tertiary enrollment), and is therefore given less weight in the computation of the IDI than the other two subindices (ITU 2013).

2. For example, smartphone bullying is a new form of bullying in Korean schools. The bullies force bullied students to sign up for subscriptions that cost around $100 a month. The bullies take over the phone's data connection to enjoy the services, because they cannot afford these kinds of high fees. Previously school bullying was not directly related to new technologies; however, the current form of school bullying is conspicuous with the growth of smartphones. The digital divide in smartphones has resulted not only in economic disparities but also in cyberbullying.

CHAPTER 7

1. As Leslie Shade (2007, 179) points out, "The mobile telephone is increasingly designed and marketed to appeal to women and female teenagers. Whether through the creation of features and accessories (ringtones, wallpaper, faceplates, camera phone, wireless synergies) or through branded phones of high-end fashion designers, the design of the mobile reflects a distinctive feminization.

CHAPTER 8

1. The country's major mobile carriers, such as SK telecom, LG UPlus, and KT, launched LTE networks in 2011. Korean wireless telecommunications service providers have continued to provide advanced LTE services, from LTE to LTE-A (2013), and again to 3 Band LTE (2014). The major difference is speed. For example, according to their service plan, in order to download 1GB-size visual images, LTE takes 110 seconds, and LTE-A takes 37 seconds; however, it only takes 28 seconds for the 3 Band LTE-A service (*Wireless* 2015).

2. For example, if 10,000 users spend 10,000 hours on Kakao Talk, the average time they spent is one hour, but if 1,000 users spend 2,000 hours on Daum Cafe, the average time they stay on is 2 hours.

References

Abeele, Mariek M. P. Vanden 2016. "Mobile Youth Culture: A Conceptual Development." *Mobile Media and Communication* 4 (1): 85–101.

Acuna, Abel. 2013. "Why Is Mobile Gaming So Popular in South Korea?" *Venture Beat.* http://venturebeat.com/2013/10/19/why-is-mobile-gaming-so-popular-in-south-korea/. Accessed June 1, 2014.

AFP. 2014. "Smartphone Sales Top a Billion, Samsung Remains Biggest Vendor." January 29. http://www.thestar.com.my/business/business-news/2014/01/29/smartphone-sales-top-a-billion-samsung-remains-worlds-biggest-vendor-saw-growth-of-429-last-year/. Accessed June 4, 2016.

Agar, Jon. 2003. *Constant Touch: A Global History of the Mobile Phone.* Duxford, Cambridge: Icon Books.

Amsden, Alice. 1989. *Asia's Next Giant: South Korea and Late Industrialization.* New York: Oxford University Press.

Anable, Aubrey. 2013. "Casual Games, Time Management, and the Work of Affect." *Journal of Gender, New Media and Technology* 2. http://adanewmedia.org/2013/06/issue2-anable/. Accessed November 21, 2015.

An Hyo-mun, and Ku Ki-sŏng. 2014. "Hwich'ŏng kŏri nŭn naebigeisŏn, chŏnggŭl esŏ saranamŭlkka" (Could the Navigation Industry Survive in the Jungle). *Auto Times.* April 30. http://autotimes.hankyung.com/apps/news.sub_view?popup=0&nid=01&c1=01&c2=&c3=&nkey=201404291433331. Accessed June 2, 2014.

App Annie. 2014. *App Annie Index: Market Q3 2014.* Market Research Report.

Appadurai, Arjun. 1996. *Modernity at Large: Cultural Dimensions of Globalization.* Minneapolis: University of Minnesota Press.

Apple. 2007a. "Apple and T-Mobile Announce Exclusive Partnership for iPhone in Germany." Press release. September 19.

Apple. 2007b. "Apple Chooses O2 as Exclusive Carrier for iPhone in UK." Press release. September 18.

Apple. 2007c. "Apple Chooses Cingular as Exclusive US Carrier for Its Revolutionary iPhone." Press Release. January 9.

Apple. 2014. "Getting Started with In-App Purchase on iOS and OS X." https://developer.apple.com/in-app-purchase/In-App-Purchase-Guidelines.pdf. Accessed February 6, 2015.

Apple. 2015. "Creating Jobs through Innovation." January 8. http://www.apple.com/about/job-creation/. Accessed November 21, 2015.

Asian Correspondent. 2010. "Why Has Google Failed in Korea?" April 29. http://asian correspondent.com/2010/04/why-has-google-failed-in-korea/. Accessed December 7, 2015.

Associated Press. 2013. "Smartphone Wars: Samsung Profit Soars as Galaxy Outsells iPhone for 4th Straight Quarter." January 25.

Associated Press. 2014. "S. Korea Rumor Crackdown Threatens Popular Message App Kakao Talk, Growing Tech Industry Star." October 5.

AT&T Archives. 2011. "Testing the First Public Cell Phone Network." http://techchannel. att.com/play-video.cfm/2011/6/13/AT&T-Archives-AMPS:-coming-of-age. Accessed January 5, 2014.

AT&T Archives. 2013. "Advanced Mobile Phone Service (AMPS)." AT&T Archives. http://techchannel.att.com/play-video.cfm/2014/2/3/ATT-Archives-Advanced-Mobile-Phone-Service-AMPS. Accessed January 5, 2014.

Bae, Cheol Gi. 2014. "The Transformation of Korean Wireless Telecommunications Policy: The State, Transnational Forces, Businesses, and Networked Users." Dissertation. University of Illinois.

Baek, B. Y. 2013. "Pushing the Restart Button." *Korea Times*. December 31.

Baguley, Richard. 2013. "The Gadget We Miss: The Nokia 9000 Communicator. Nokia's First Smartphone was a Ground-Breaking Gadget for the Traveler." August 1. https://medium.com/people-gadgets/ef8e8c7047ae. Accessed February 5, 2014.

Bainbridge, Jane. 2013. "Last-Mover Advantage: When Brands Should Avoid Taking the Lead." *Campaign*. January 24. http://www.marketingmagazine.co.uk/article/1167 648/last-mover-advantage-when-brands-avoid-taking-lead#6izPc3OSPd TCOFG4.99. Accessed March 1, 2014.

Behrooz. 2012. "Smartphones: Interaction Technology Design in Context." http://itdic smu.blogspot.ca/2013/01/smartphones.html. Accessed February 2, 2015.

BellSouth-IBM Simon. 1994. "Retro Computing Reviewer—BellSouth IBM Simon." http://www.retrocom.com/bellsouth_ibm_simon.htm. Accessed February 6, 2014.

Benkler, Yochai. 2006. *The Wealth of Networks: How Social Production Transforms Markets and Freedom*. New Haven: Yale University Press.

Bijker, Wiebe, Thomas Hughes, and Trevor Pinch. 2012. *The Social Construction of Technological Systems*. Anniversary edition. Cambridge, MA: MIT Press.

Bischoff, Paul. 2014. "Korean Mobile Game Maker 4:33 Creative Lab Attracts Big Investment from Tencent and Line." *Game in Asia*. November 13. https://www.techinasia. com/korean-mobile-game-maker-433-creative-lab-attracts-big-investment-ten cent-line/. Accessed January 6, 2015.

Bolin, Goran. 2012. "Personal Media in the Digital Economy." In *Moving Data: The iPhone and the Future of Media*, ed. Pelle Sinckars and Patrick Vonderau, 91–103. New York: Columbia University Press.

Boston Globe. 2014. "History of the Cellphone." January 9, A10.

boyd, danah. 2011. "Social Network Sites as Networked Publics: Affordances, Dynamics, and Implications." In *A Networked Self: Identity, Community, and Culture on Social Network Sites*, ed. Zizi Papacharissi, 39–58. London: Routledge.

Boyd-Barrett, Oliver. 2006. "Cyberspace, globalization and empire." *Global Media and Communication* 2 (1): 21–41.

Boyd-Barrett, Oliver, and S. Xie. 2008. "Al-Jazeera, Phoenix Satellite Television and the

Return of the State: Case Studies in Market Liberalization, Public Sphere and Media Imperialism." *International Journal of Communication* 2: 206–22.

Bradner, Erin. 2011. "Are You an Innovation Giant?" July 11. http://dux.typepad.com/dux/2011/07/are-you-an-innovation-giant-.html. Accessed March 1, 2014.

Brey, Philip. 1997. "Social Constructivism for Philosophers of Technology: A Shopper's Guide." *Society for Philosophy and Technology* 2 (3–4). http://www.utwente.nl/bms/wijsb/organization/brey/Publicaties_Brey/Brey_1997_Social-Constructivism_PoT.pdf. Accessed March 3, 2014.

Brownell, Claire. 2015. "CRTC Relaxes Content Rules to Help Canadian TV Broadcasters Compete with Digital Media. *Financial Post.* March 12. http://business.financialpost.com/fp-tech-desk/crtc-relaxes-quotas-on-canadian-content-for-tv-broadcasters?__lsa=6127-5edd. Accessed June 6, 2016.

Business Services Industry. 1993. "Bellsouth, IBM Unveil Personal Communicator Phone." *Mobile Phone News* 8: 1–3.

Call, Joshua, Katie Whitlock, and Gerald Voorhees. 2012. "From Dungeons to Digital Denizens." In *Dungeons, Dragons, and Digital Denizens: The Digital Role-Playing Game,* ed. Gerala Voorhees, Joshua Call, and Katie Whitlock, 11–24. London: Continuum.

Campbell, Scott, and Tracy Russo. 2003. "The Social Construction of Mobile Telephony: An Application of the Social Influence Model to Perceptions and Uses of Mobile Phones within Personal Communication Networks." *Communication Monographs* 70 (4): 317–34.

Cammaerts, Bart, Leo Van Audenhove, Gert Nulens, and Caroline Pauwels, eds. 2003. *Beyond the Digital Divide: Reducing Exclusion and Fostering Inclusion.* Brussels: VUB Press.

Caoli, Eric. 2011. "This Week in Korean Online Gaming News: From NCsoft to Mobile Game Ratings." *Gamasutra.* July 8. http://gamasutra.com/view/news/35753/This_Week_In_Korean_Online_Gaming_News_From_NCsoft_To_Mobile_Game_Ratings.php. Accessed May 1, 2014.

Castells, Manuel. 1996. *The Information Age: Economy, Society and Culture.* Vol. 1: *The Rise of the Network Society.* Oxford: Blackwell.

Castells, Manuel. 1997. *The Information Age: Economy, Society and Culture.* Vol. 2: *The Power of Identity.* Oxford: Blackwell.

Castells, Manuel. 2001. *The Internet Galaxy: Reflections on the Internet, Business, and Society.* New York: Oxford University Press.

Castells, Manuel. 2007. "Communication, Power and Counter-power in the Network Society." *International Journal of Communication* 1: 238–66.

Castells, Manuel. 2009. *Communication Power.* London: Oxford University Press.

Castells, Manuel. 2010. *The Rise of the Network Society.* 2nd ed. Oxford: Wiley-Blackwell.

Center for Innovation Management Studies. 2013. "The Simon Personal Communicator." http://cims.ncsu.edu/the-simon-personal-communicator/. Accessed May 1, 2014.

Chae, H. M. 1997. "Info Superhighway to Be Laid by 2010." *Korea Times.* July 29, 8.

Chambers, John. 2013. "Transforming I.T. for the Application Economy." *Cisco Blog.* November 6. http://blogs.cisco.com/news/transforming-i-t-for-the-application-economy/. Accessed March 3, 2014.

Chan, Dean. 2008. "Convergence, Connectivity, and the Case of Japanese Mobile Gaming." *Games and Culture* 3 (1): 13–25.

Chandler, Daniel. 1996. "Shaping and Being shaped: Engaging with Media." *Computer-Mediated Communication Magazine* 3 (2). http://www.december.com/cmc/mag/1996/feb/chandler.html. Accessed January 2014.

Chang Yun-hŭi. 2015. "Mobail mesinjŏ, 'k'ak'aot'ok' chŏnha" (Kakao Talk Monopolizes the Mobile Messenger Market). *Nyusisŭ* (*Newsis*). May 10. http://www.newsis.com/ar_detail/view.html?ar_id=NISX20150510_0013652118&cID=10402&pID=10400. Accessed November 26, 2015.

Cheng, Chih-Wen. 2012. "The System and Self-reference of the App Economy: The Case of Angry Birds." *Westminster Papers in Communication and Culture* 9 (1): 47–66.

Cheng, Jonathan. 2014. "South Korean Startup Woowa Brothers Attracts $36 Million in Funding Round." *Wall Street Journal*. November 26. http://online.wsj.com/articles/south-korea-startup-woowa-brothers-attracts-36-million-in-funding-round-1417037590?tesla. Accessed November 30, 2014.

Chin, Dae-Jin, and Myung-Hwan Rim. 2006. "IT839 Strategy: The Korean Challenge toward a Ubiquitous World." *IEEE Communications Magazine*. April, 32–38.

Chircu, Alina, and Vijay Mahajan. 2009. "Perspective: Revisiting the Digital Divide. An Analysis of Mobile Technology Depth and Service Breadth in the BRIC Countries." *Journal of Product Innovation Management* 26 (4): 455–66.

Cho, Jin-seo. 2007. "KTF in Talks to Sell iPhone in Korea." *Korea Times*. August 23. http://koreatimes.co.kr/www/news/tech/2007/08/129_8832.html

Cho Kwang-min. 2014. "Net mabŭl, 'monsŭt'ŏ kilt'ŭrigi' kilt'ŭ sisŭt'em ŏpteit'ŭ" (Netmarble Updates Monster Taming System). *Keim Tonga*. January 21. http://game.donga.com/71657/. Accessed April 2, 2015.

Cho, Mu-hyun. 2012. "Player of Puzzle Game Ani Pang Tops 20 Mil." *Korea Times*. October 12. http://www.koreatimes.co.kr/www/news/tech/2013/01/133_122101.html. Accessed March 3, 2013.

Choi, Eunjeong. 2013. "Kakao Talk, a Mobile Social Platform Pioneer." *SERI Quarterly* 6 (1): 62–69.

Choi, You Lee. 2015. "Google Turns Over the Rank in the Mobile Search Engine Sector." *Hankyung Economic Daily*. May 28. http://www.hankyung.com/news/app/news view.php?aid=201505279023g. Accessed August 6, 2016.

Chŏng Po-ra. 2013. "K'ak'aot'ok kaipcha 1 ŏk myŏng tolp'a" (Kakao Talk Users Passed 100 Million). *Pŭllot'ŏ*. http://www.bloter.net/archives/157860. Accessed December 15, 2013.

Chosun Ilbo. 2003. "SKT Speed Up in the Mobile Phone Service Market." July 31, 24.

Christensen, Christian, and Patrick Prax. 2012. "Assemblage, Adaptation and Apps: Smartphones and Mobile Gaming." *Continuum* 26 (5): 731–39.

Clarke, John. 1976. "Style." In *Resistance through Rituals: Youth Subcultures in Post-war Britain*, ed. Stuart Hall and Tony Jefferson, 147–61. 2nd ed. London: Routledge.

CNN. 2013. "10 Things South Korea Does Better Than Anywhere Else." November 27. http://www.cnn.com/2013/11/27/travel/10-things-south-korea-does-best/

CNN. 2014. "Smartphones Are Fading: Wearables Are Next." March 19. http://money.cnn.com/2014/03/19/technology/mobile/wearable-devices/. Accessed May 2, 2014.

Communities Dominate Brands. 2015. "Smartphone Wars: Q3 Scorecard—All Market Shares, Top 10 Brands, OS Platforms, Installed Base." October 30. http://communi

ties-dominate.blogs.com/brands/2015/10/smartphone-wars-q3-scorecard-all-market-shares-top-10-brands-os-platforms-installed-base.html

Compaine, Benjamin, ed. 2001. *The Digital Divide: Facing a Crisis or Creating a Myth?* Cambridge, MA: MIT Press.

Computerworld. 1983. "Features Enhanced Voice: Baby Bell Offers Smart Phones." January 31, 47.

ComScore. 2014. "ComScore Reports August 2014 US Smartphone Subscriber Market Share." October 7. http://www.comscore.com/Insights/Market-Rankings/comScore-Reports-August-2014-US-Smartphone-Subscriber-Market-Share

Conabree, D. 2001. "Ericsson Introduces the New R380e." *Mobile Magazine.* September 25. http://www.mobilemag.com/2001/09/25/ericsson-introduces-the-new-r380e. Accessed February 25, 2015.

Connelly, C. 2012. "Apple Not the First to Get Smart." *Advertiser.* November 23.

Couvering, Elizabeth Van. 2012. "Search Engines in Practice: Structure and Culture in Technical Development." In *Cultural Technologies: The Shaping of Culture in Media and Society,* ed. Goran Bolin, 118–32. London: Routledge.

Crawley, Dan. 2014. "Google Play Downloads Are 60% Higher Than on iOS App Store, but Apple Still Rules on Revenue." *Gamesbeat.* October 15. http://venturebeat.com/2014/10/15/google-play-downloads-60-percent/. Accessed February 15, 2015.

Curran, James, and Myung-Jin Park, eds. 2000. *De-westernizing Media Studies.* London: Routledge.

Cutler, Kim-Mai. 2013. "Behind South Korea's Big $65M. Mobile Gaming Merger." *Techcrunch.* October 7. http://techcrunch.com/2013/10/07/behind-south-koreas-big-65m-mobile-gaming-merger/. Accessed March 23, 2014.

Daliot-Bul, Michal. 2007. "Japan's Mobile Technoculture: The Production of a Cellular Playscape and Its Cultural Implications." *Media, Culture and Society* 29 (6): 954–71.

Damouni, Nadia, Nicole Leske, and Gerry Shih. 2014. "Lenovo to Buy Google's Motorola in China's Largest Tech Deal." Reuters. January 29. http://www.reuters.com/article/2014/01/29/us-google-lenovo-idUSBREA0S1YN20140129. Accessed February 2, 2014.

Demont-Heinrich, Christof. 2008. "The Death of Cultural Imperialism—and Power Too?" *International Communication Gazette* 70 (5): 378–94.

Daubs, Michael S., and Vincent Manzerolle. 2016. "App-centric Mobile Media and Commoditization: Implications for the Future of the Open Web." *Mobile Media and Communication* 4 (1): 52–68.

DiMaggio, Paul, Eszter Hargittai, Russell Neuman, and John Robinson. 2001. "Social Implications of the Internet." *Annual Review of Sociology* 27: 307–36.

Distimo. 2013. "2013 Year in Review." December.

Dong-a Ilbo (Tonga ilbo). 2014. "Samsŏng mobail mesinjŏ 'Ch'aedon' sŏbisŭ chongnyo" (Samsung Electronic Co. Ends Its Mobile Messenger Business ChatOn). November 24. http://news.donga.com/Main/3/all/20141124/68131968/1. Accessed May 1, 2015.

Dou, Eva. 2013. "Apple Shifts Supply Chain Away from Foxconn to Pegatron." *Wall Street Journal.* May 29. http://online.wsj.com/news/articles/SB10001424127887323855804578511122734340726. Accessed December 4, 2014.

Dredge, Stuart. 2013. "Clash of Clans Is 2013's Most Lucrative Gaming App, Data Shows." *Guardian.* December 18. http://www.theguardian.com/technology/2013/dec/18/android-ios-app-revenues-research. Accessed November 2, 2014.

du Gay, Paul, Stuart Hall, Linda Janes, Hugh Markay, and Keith Negus. 1997. *Doing Cultural Studies: The Story of the Sony Walkman.* London: Sage.

Dunnewijk, Theo, and Staffan Hulten. 2005. "A Brief History of Mobile Phones in Europe." *Telematics and Informatics* 27: 164–79.

Dyer-Witheford, Nick. 2014. "App Worker." In *The Imaginary App*, ed. Paul Miller and Svitlana Matviyenko, 127–41. Cambridge, MA: MIT Press.

Economist. 2014. "Daum and Kakao Talk Merge: Getting the Message." May 31. http://www.economist.com/news/business/21603035-latest-tie-up-between-messaging-apps-and-broader-online-firms-getting-message. Accessed June 8, 2014.

Electronics and Telecommunications Research Institute. 2001. *Report on the Composition and Operation of Wireless Internet Standardization Forum.* Daejeon: ETRI.

EMarketer. 2013. "With Mature US Online Population, Small Gains for Email, Search Usage." March 4. http://www.emarketer.com/Article/With-Mature-US-Online-Population-Small-Gains-Email-Search-Usage/1009704#sthash.7f8HHumE.dpuf

Engel, Joel S. 2008. "The Early History of Cellular Telephone." *IEEE Communication Magazine.* August, 27–29.

Federal Communications Commission. 2005. "The Quality That Made Radio Popular." http://transition.fcc.gov/omd/history/radio/quality.html

Fehske, Albrecht, Gerhard Fettweis, Jens Malmodin, and Gergely Biczok. 2011. "The Global Footprint of Mobile Communications: The Ecological and Economic Perspective." *IEEE Communications Magazine* 49 (8): 55–62.

Fingas, Jon. 2013. "Strategy Analytics: Android Claimed 70 Percent of World Smartphone Share in Q4 201229 January." Strategy Analytics. Press release.

Fischer, Claude. 1992. *America Calling: A Social History of the Telephone to 1940.* Berkeley: University of California Press.

Fiske, John. 1987. *Television Culture.* London: Routledge.

Flanagin, Andrew, Craig Flanagin, and Jon Flanagin. 2010. "Technical Code and the Social Construction of the Internet." *New Media and Society* 12 (2): 179–96.

Forbes. 2007. "SK Telecom Acquiring Control of Hanaro for $1.2B." December 3. http://www.forbes.com/2007/12/03/sk-hanaro-telecom-markets-equity-cx_vk_1203markets2.html. Accessed February 14, 2014.

Fortunati, Leopoldina. 2012. "Mobile Communication and the Fourth Communicative Revolution." In *Mobile Communication and Greater China*, ed. Chu, Rodney, Leopoldina Fortunati, Pui-lam Law, and Shanhua Yang, 49–63. London: Routledge.

Frenkiel, Richard. 2010. "Creating Cellular: A History of the AMPS Project (1971–1983)." *IEEE Communications Magazine.* September, 14–24.

Friedman, Milton. 2002. *Capitalism and Freedom.* Fortieth Anniversary Edition. Chicago: University of Chicago Press.

Friedman, Thomas. 2005. *The World Is Flat: A Brief History of the 21st Century.* New York: Farrar, Straus and Giroux.

Frier, Sarah. 2014. "Facebook $22 Billion WhatsApp Deal Buys $10 Million on Sales." *Bloomberg.* October 30. http://www.bloomberg.com/news/2014-10-28/facebook-s-22-billion-whatsapp-deal-buys-10-million-in-sales.html. Accessed December 1, 2014.

Fuchs, Christian. 2014. *Social Media: A Critical Introduction.* London: Sage.

Garcia de la Garza, Alejandro. 2013. "From Utopia to Dystopia: Technology, Society and What We Can Do about it." *Open Security.* December 20. https://www.opendemoc

racy.net/opensecurity/alejandro-garcia-de-la-garza/from-utopia-to-dystopia-tech nology-society-and-what-we-can. Accessed December 27, 2013.

Giddens, Anthony. 1999. "Runaway World: 1999 Reith Lecture: Globalization." http:// www.bbc.co.uk/radio4/reith1999/lecture1.shtml

Gillespie, Tarleton. 2010. "The Politics of Platforms." *New Media and Society* 12 (3): 347–64.

Goff, David. 2013. "A History of the Social Media Industries." In *The Social Media Industries*, ed. Alan Albarran, 16–45. London: Routledge.

Goggin, Gerald. 2007. "Introduction: Mobile Phone Cultures." *Continuum* 21 (2): 133–35.

Goggin, Gerald. 2009. "Adapting the Mobile Phone: The iPhone and Its Consumption." *Continuum* 23 (2): 231–44.

Goggin, Gerald. 2011a. "Ubiquitous Apps: Politics of Openness in Global Mobile Cultures." *Digital Creativity* 22 (3): 148–59.

Goggin, Gerald. 2011b. *Global Mobile Media*. London: Sage.

Goldman, David. 2015. "This Is the Apple of China." *CNN Money*. January 5. http:// money.cnn.com/2015/01/05/technology/mobile/xiaomi-china-apple/index. html?iid=HP_LN. Accessed February 3, 2015.

Goldsmith, Jack, and Tim Wu. 2006. *Who Controls the Internet: Illusion of a Borderless World*. New York: Oxford University Press.

Gomery, Douglas. 1996. "The Hollywood Studio System." In *The Oxford History of World Cinema*, ed. Geoffrey Nowell-Smith, 43–52. New York: Oxford University Press.

Gonzalez, Jorge. 2000. "Cultural Fronts: Towards a Dialogical Understanding of Contemporary cultures." In *Culture in the Communication Age*, ed. James Lull, 106–31. New York: Routledge.

Google. 2006. "Google to Acquire YouTube for $1.65 Billion in Stock." Press release. October 9.

Google. 2014. "Development Distribution Agreement." https://play.google.com/about/ developer-distribution-agreement.html. Accessed March 2, 2015.

Gunkel, David. 2003. "Second Thoughts: Toward a Critique of the Digital Divide." *New Media and Society* 5 (4): 499–522.

Grubb, Jeff. 2014. "Game-Distribution Platform (and Messaging App) Kakao Has Driven 500 Million Downloads on iOS and Android." *VentureBeat*. May 14. http://venture-beat.com/2014/05/14/game-distribution-platform-and-messaging-app-kakao-has-driven-500-million-downloads-on-ios-and-android/. Accessed February 23, 2015.

Ha, Peter. 2010. "All-Time 100 Gadgets." *Time*. October 25. http://content.time.com/ time/specials/packages/article/0,28804,2023689_2023708_2023604,00.html. Accessed February 24, 2014.

Halevy, R. 2009. The History of RIM & the BlackBerry Smartphone, Part 3: The Evolution Of Color. March 16. Research in Motion first released its GSM BlackBerry 6210 in 2003, and later released the Blackberry 7730. Accessed June 4, 2016.

Hardt, Michael, and Antonio Negri. 2000. *Empire*. Cambridge, MA: Harvard University Press.

Hargittai, Eszter, and Su Jung Kim. 2012. "The Prevalence of Smartphone Use among a Wired Group of Young Adults." Working paper, Northwestern University. http:// www.ipr.northwestern.edu/publications/papers/2011/ipr-wp-11-01.html. Accessed October 24, 2015.

Harjani, Ansuya. 2014. "What's the Future of Smartphones? Think Personal Assistant." CNBC. April 3. http://www.cnbc.com/id/101521265

Hart-Landsberg, Martin. 1993. *The Rush to Development: Economic Change and Political Struggle in South Korea*. New York: Monthly Review Press.

Hart-Landsberg, Martin. 2013. *Capitalist Globalization: Consequences, Resistance, and Alternatives*. New York: Monthly Review Press.

Harvey, David. 2007. "In What Ways Is the New Imperialism Realty New." *Historical Materialism* 15 (3): 57–70.

Havens, Timothy, and Amanda Lotz. 2012. *Understanding Media Industries*. New York: Oxford University Press.

Hartung, Adam. 2014. "Three Smart Lessons from Facebook's Purchase of WhatsApp." *Financial Times*. February 24. http://www.forbes.com/sites/adamhartung/2014/02/24/zuckerbergs-3-smart-leadership-lessons-from-facebook-buying-whatsapp/

Heo, Uk, and Sunwoong Kim. 2000. "Financial Crisis in South Korea: Failure of the Government-Led Development Paradigm." *Asian Survey* 40 (3): 492–507.

Herring, Susan C. 2003. "Computer-Mediated Discourse." In *Handbook of Discourse Analysis*, Ed. Deborah Tannen, Deborah Shiffrin, and Heidi Hamilton, 612–34. Oxford: Wiley-Blackwell.

Hjorth, Larissa. 2006. "Playing at Being Mobile: Gaming and Cute Culture in South Korea." *Fibreculture Journal* 8. http://eight.fibreculturejournal.org/fcj-052-playing-at-being-mobile-gaming-and-cute-culture-in-south-korea/. Accessed March 4, 2015.

Hjorth, Larissa. 2007a. "The Game of Being Mobile: One Media History of Gaming and Mobile Technologies in Asia-Pacific." *Convergence* 13 (4): 212–21.

Hjorth, Larissa. 2007b. "Home and Away: A Case Study of the Use of Cyworld Mini-hompy by Korean Students Studying in Australia." *Asian Studies Review* 31 (4): 397–407.

Hjorth, Larissa. 2009. *Mobile Media in the Asia-Pacific: Gender and the Art of Being Mobile*. London: Routledge.

Hjorth, Larissa. 2011a. "Mobile@game Cultures: The Place of Urban Mobile Gaming." *Convergence* 17 (4): 357–71.

Hjorth, Larissa. 2011b. *Games and Gaming: An Introduction to New Media*. Oxford: Berg.

Hjorth, Larissa. 2012. "iPersonal: A Case Study of the Politics of the Personal." In *Studying Mobile Media: Cultural Technologies, Mobile Communication, and the iPhone*, ed. Larissa Hjorth, Jean Burgess, and Ingrid Richardson, 190–212. London: Routledge.

Hjorth, Larissa, Jean Burgess, and Ingrid Richardson, eds. 2012. *Studying Mobile Media: Cultural Technologies, Mobile Communication, and the iPhone*. London: Routledge.

Hjorth, Larissa, and Dean Chan. 2009. "Locating the Game: Gaming Cultures in/and the Asia-Pacific." In *Gaming Cultures and Place in Asia-Pacific*, ed. Larissa Hjorth and Dean Chan, 1–14. New York: Routledge.

Horwitz, Robert. 1986. "For Whom the Bell Tolls: Causes and Consequences of the AT&T Divestiture." *Critical Studies in Mass Communication* 3 (2): 119–54.

Hozy, B. 1985. "Integrated Voice Data Terminals: Telephone and More at Touch of Dial." *Financial Post*. February 2, S3.

Hughes, Thomas. 1994. "Technological Momentum." In *Does Technology Drive History: The Dilemma of Technological Determinism*, ed. Merritt Roe Smith and Leo Marx, 99–113. Cambridge, MA: MIT Press.

Huh, J. K. 2011. "Interview with Jae Bum Lee of Kakao Inc." *Hankook Ilbo*. February 10, 17.

Humphreys, Lee. 2013. "Mobile Social Media: Future Challenges and Opportunities." *Mobile Media and Communication* 1 (1): 20–25.

Hung, Quoc. 2009. "Samsung Opens US$700 Million Phone Plant." *Saigon Times.* October 30. http://english.thesaigontimes.vn/7173/Samsung-opens-US$700-million-phone-plant.html. Accessed June 6, 2016.

Husted, B. 1991. "Business Report: On Technology; Civil War Care Was Hard to Swallow." *Atlanta Journal and Constitution.* October 31, 2.

Hwang, J. H. 2008. "Government's Plan to Review Mandatory Policy for WIPI Igniting Debate." *Korea IT News.* August 28. http://english.etnews.com/communica tion/2389925_1300.html

Hwang, You Sun, and Namkee Park. 2013. "Digital Divide in Social Networking Sites." *International Journal of Mobile Communications* 11 (5): 446–64.

IBM Corporate Archives. N.d. *1990–1995 IBM Highlights.* Armonk, NY: IBM.

Im Wŏn-gi. 2013. "Hanguk ŭi sŭt'at'ŭŏp sijŭn 2-(1) Sŏndei T'oju, Aenip'ang kŭ ihu". *Int'ŏnet Insaidŭ.* April 9. http://limwonki.com/571

Independent. 2009. "iPhone's Debut in S. Korea Means Paradigm Shift." November 27. http://www.independent.co.uk/life-style/gadgets-and-tech/news/iphones-debut-in-skorea-means-paradigm-shift-1830631.html

Independent. 2014. "Smartphone at 20: IBM Simon Becomes Museum Exhibit." August 16. http://www.independent.co.uk/life-style/gadgets-and-tech/news/the-smartphone-at-20-ibm-simon-becomes-museum-exhibit-9673528.html. Accessed October 2, 2014.

International Data Corporation. 2010. "Nokia Owned the Global Smartphone Space in 2009." Press release. February 5.

International Data Corporation. 2012. "Worldwide Mobile Phone Market Maintains Its Growth Trajectory." Press release. February 1.

International Data Corporation. 2013. "Strong Demand for Smartphones and Heated Vendor Competition Characterize the Worldwide Mobile Phone Market at the End of 2012." Press release. January 24.

International Data Corporation. 2014a. Worldwide Mobile Phone Tracker. January 27. http://www.idc.com/tracker/showproductinfo.jsp?prod_id=37. Accessed December 1, 2015.

International Data Corporation. 2014b. "Smartphone Vendor Market Share, Q2 2014." http://www.idc.com/prodserv/smartphone-market-share.jsp. Accessed December 1, 2015.

International Data Corporation. 2016. "Smartphone Shipments Flat for the First Time; Samsung Widens Lead over Apple in Q1 2016." http://venturebeat.com/2016/04/27/idc-smartphone-shipments-flat-for-the-first-time-samsung-widens-lead-over-ap ple-in-q1-2016/. Accessed June 2, 2016.

Internet Trend. 2014. "Market Share of Search Engines." http://www.internettrend.co.kr/trendForward.tsp

International Telecommunication Union. 2013. *Measuring the Information Society.* Geneva: ITU.

Ito, Mizuko, and Daisuke Okabe. 2005. "Intimate Connections: Contextualizing Japanese Youth and Mobile Messaging." In *The Inside Text: Social, Cultural and Design Perspectives on SMS*, ed. R. Harper, L. Palen, and A. Taylor, 127–45. Berlin: Springer-Verlag.

Ito, Mizuko, Daisuke Okabe, and Misa Matsuda. 2005. *Personal, Portable, Pedestrian: Mobile Phones in Japanese Life*. Cambridge, MA: MIT Press.

Iwabuchi, Koichi. 2010. "De-westernization and the Governance of Global Cultural Connectivity: A Dialogic Approach to East Asian Media cultures." *Postcolonial Studies* 13 (4): 403–19.

Jenkins, Henry, Sam Ford, and Joshua Green. 2013. *Spreadable Media: Creating Value and Meaning in a Networked Culture*. New York: New York University Press.

Jenkins, Henry. 2006. *Convergence Culture: Where Old and New Media Collide*. New York: New York University Press.

Jensen, Klaus Bruhn. 2013. "What's Mobile in Mobile Communication." *Mobile Media and Communication* 1 (1): 26–31.

Jho, Whasun. 2007. "Global Political Economy of Technology Standardization: A Case of the Korean Mobile Telecommunications Market." *Telecommunications Policy* 31: 124–38.

Jho, Whasun. 2013. *Building Telecom Markets: Evolution of Governance in the Korean Mobile Telecommunication Market*. Berlin: Springer.

Ji, Pan, and Marko Skoric. 2013. "Gender and Social Resources: Digital Divides of Social Network Sites and Mobile Phone Use in Singapore." *Chinese Journal of Communication* 6 (2): 221–39.

Jin, Dal Yong. 2007. "Reinterpretation of Cultural Imperialism: Emerging Domestic Market vs. Continuing U.S. Dominance." *Media, Culture and Society* 29 (5): 753–71.

Jin, Dal Yong. 2010. *Korea's Online Gaming Empire*. Cambridge, MA: MIT Press.

Jin, Dal Yong. 2011. *Hands on/Hands off: The Korean State and the Market Liberalization of the Communication Industry*. New York: Hampton Press.

Jin, Dal Yong. 2014. "Construction of the App Economy in the Networked Korean Society." In *The Imaginary App*, ed. Paul Miller and Svitlana Matviyenko, 163–78. Cambridge, MA: MIT Press.

Jin, Dal Yong. 2015. *Digital Platforms, Imperialism and Political Culture*. London: Routledge.

Jin, Dal Yong. 2016. *New Korean Wave: Transnational Popular Culture in the Age of Social Media*. Urbana: University of Illinois Press.

Jin, Dal Yong, Florence Chee, and Seah Kim. 2015. "Transformative Mobile Game Culture: Socio-cultural Analysis of the Korean Mobile Gaming in the Smartphone Era." *International Journal of Cultural Studies* 18 (4): 413–29.

Jin, Dal Yong, and Dong-Hoo Lee. 2012. "The Birth of East Asia: Cultural Regionalization through Co-production Strategies." *Spectator* 32 (2): 26–40.

Jin, Hyun-joo. 2008a. "Apple, Nokia Eye Korean Market." *Korea Herald*. August 21.

Jin, Hyun-joo. 2008b. "Korea Removes Hurdle for iPhone Sale." *Korea Herald*. December 11.

Johnson, Bradley, and K. Fitzgerald. 1994. "BellSouth Puts Smarts in Simon Cellular Phone." *Advertising Age*. February 7, 8.

Jones, Steve, Veronika Karnowski, Richard Ling, and Thilo von Pape. 2013. "Welcome to *Mobile Media and Communication*." *Mobile Media and Communication* 1 (1): 3–7.

Jung, Man-Won. 2010. "South Korea's Future in Mobile and Wireless." *Asia-Pacific III*. 8–9.

Juul, Jesper, ed. 2005. "Where the Action Is." *Game Studies* 5 (1). http://www.gamestudies.org/0501/. Accessed March 20, 2013.

Juul, Jesper. 2010. *A Casual Revolution: Reinventing Video Games and Their Players.* Cambridge, MA: MIT Press.

Ju Ŭn-a. 2014. "Kukŭl ŭi 'twit'ongsu ttaerigi', it'ongsa ŭi chasŭngjabak" (Google Proposed a New Plan in App Service Fees). *Business Post.* January 17. http://www.businesspost. co.kr/news/articleView.html?idxno=217. Accessed March 23, 2014.

Kane, Yukari, and Ben Worthen. 2010. "As iPhone Goes Global, App Makers Follow." *Wall Street Journal.* April 29. http://online.wsj.com/news/articles/SB1000142405274 8703648304575212461802126530. Accessed September 29, 2013.

Katz, James, and Satomi Sugiyama. 2006. "Mobile Phones as Fashion Statements: Evidence from Student Surveys in the US and Japan." *New Media and Society* 8 (2): 321–37.

Kerner, Sean. 2013. "Chambers: Get Ready for the App Economy Now." Internetnews. com. October 3. http://www.internetnews.com/infra/chambers-get-ready-for-the-app-economy-now.html. Accessed April 25, 2014.

Kim Chae-sŏp. 2014. "Aip'on-6 ch'ulsi 'D-1' . . . it'ong 3-sa 300-man 'aip'onppa' chapki chŏnjaeng" (iPhone Sale D-1: Three Wireless Telecommunications Service Providers Try to Catch 3 Million Apple Mania). *Hangyŏre Sinmun.* October 31. http://www. hani.co.kr/arti/economy/it/662019.html. Accessed November 28, 2014.

Kim, D. H. 2002. "Korea Becomes Global Leader in Broadband Internet." *Korea Times.* November 6.

Kim, Gwangseok. 2011. "iPhone Effect: A Critical Analysis of Discourses on Science and Technology in South Korea." Paper presented at the International Communication Association Conference, Boston, MA, May 25.

Kim Ho-gi, Sin Ki-uk, Go Dong-hyŏn, Yi Sŭng-hun. 2011. "Sŭmat'ŭp'on sidae ŭi mobail tibaidŭ" (Mobile Divide in the Age of Smartphone). KT Kyŏngje Kyŏngyŏng Yŏn'guso (KT Business and Economics Research Center). http://www.digieco.co.kr/ KTFront/index.action. Accessed March 26, 2014.

Kim, Hyung-eun. 2011. "Big Bang of Mobile Games." *Joongang Daily.* March 17. http:// koreajoongangdaily.joins.com/news/article/article.aspx?aid=2936279

Kim, Pyungho. 2011. "The Apple iPhone Shock in Korea." *Information Society* 27 (4): 261–68.

Kim Sang-yun. 2014 "'Aenip'ang2 ŭi him': Sŏndei T'oju sasang ch'oedae silchŏk" (The Power of Anypang 2—Sunday Toz Enjoyed the Historical Growth). *Jungang Ilbo* (*Joongang Daily*). May 8. http://news.joins.com/article/14623525. Accessed June 7, 2016.

Kim Suk. 2014. "Kungnae wept'un sanŏp i hallyu chisok e mich'inŭn yŏnghyang" (The Influence of the Domestic Webtoon Market to the Continuation of the Korean Wave). *K'ok'a p'ok'ŏsŭ* (*KOCCA Focus*) 86. Seoul: Han'guk k'ont'ench'ŭ chinhŭngwŏn (Korea Creative Content Agency).

Kim, Sung-Young. 2012. "The Politics of Technological Upgrading in South Korea: How Government and Business Challenged the Might of Qualcomm." *New Political Economy* 17 (3): 293–312.

Kim, T. G. 2004. "Korea, US Compromise over Internet platform." *Korea Times.* April 23.

Kim, Tong-hyung. 2004. "UN Denounces Korea's Internet Platform." *Korea Times.* February 26.

Kim, Tong-hyung. 2008. "Goodbye WIPI, Hello iPhone." *Korea Times.* December 10. http://www.koreatimes.co.kr/www/news/tech/2012/04/133_35873.html. Accessed November 15, 2013.

Kim, Tong-hyung. 2009a. "KT-KTF Merger Given Green Light." *Korea Times*. March 18.

Kim, Tong-hyung. 2009b. "iPhone Has Samsung, LG Sweating." *Korea Times*. December 2. http://koreatimes.co.kr/www/news/biz/2009/12/602_56574.html. Accessed March 1, 2014.

Kim, Yang-jin. 2014. "Kugŭl, kungnae mobail kŏmsaek sijang kŭpsok chamsik" (Mobile Search Engine in Korea). *Sŏul Sinmun*. March 8.

Kim, Yun Tae. 1999. "Neoliberalism and the Decline of the Developmental State." *Journal of Contemporary Asia* 29 (4): 441–61.

King, Jomilah. 2011. "How Big Telecom Used Smartphones to Create a New Digital Divide." Colorlines.com. December 6. http://colorlines.com/archives/2011/12/the_new_digital_divide_two_separate_but_unequal_internets.html. Accessed March 3, 2013.

Kline, Steven, Nick Dyer-Witheford, and Greig de Peuter. 2003. *Digital Play: The Interaction of Technology, Culture, and Marketing*. Montreal: McGill-Queen's University Press.

Klemens, Guy. 2010. *The Cellphone: The History and Technology of the Gadget That Changed the World*. London: McFarland.

Kohiyama, Kenji. 2005. "A Decade in the Development of Mobile Communications in Japan (1993–2002)." In *Personal, Portable, Pedestrian: Mobile Phones in Japanese Life*, ed. Ito Mizuko, Daisuke Okabe, and Misa Matsuda, 61–74. Cambridge, MA: MIT Press.

Korea Communications Commission. 2010. *Project for the Development of Wireless Internet*. http:///www/kcc.go.kr/tsi/etc/search/ASC_integrationsearch.jsp?page=P100 10000. Accessed June 27, 2013.

Korea Communications Commission. 2013. *Market Share of Mobile Telecommunications in January 2013*. Seoul: KCC.

Korea Herald. 2012. "Mobile Game Spices Up Working Women's Lives." November 20. http://www.koreaherald.com/view.php?ud=20121205000752. Accessed June 7, 2013.

Korea Internet and Security Agency (Han'guk Int'ŏnet Chinhŭngwŏn). 2013. *2013-nyŏn sŭmat'ŭp'on iyong silt'ae chosa (Smartphone Use Report of 2013)*. Seoul: Han'guk Int'ŏnet Chinhŭngwŏn.

Korea Internet and Security Agency. 2014. *2014-nyŏn sŭmat'ŭp'on iyong silt'ae chosa (Annual Survey on the Internet Usage of 2014)*. Seoul: Han'guk Int'ŏnet Chinhŭngwŏn.

Korea Mobile Internet Business Association (Han'guk Musŏn Int'ŏnet Sanŏp Yŏnhaphoe). 2014. *2013-nyŏn Taehan Min'guk musŏn int'ŏnet sanŏp hyŏnhwang (Korea Mobile Internet Industry Report 2013)*. Seoul: Han'guk Musŏn Int'ŏnet Sanŏp Yŏnhaphoe.

Korea Press Foundation (Han'guk Ŏllon Chinhŭng Chaedan). 2014. *2013 Sinmunsa Chaemu Punsŏk (2013 Newspaper Financial Report)*. Seoul: Han'guk Ŏllon Chinhŭng Chaedan.

Korea Telecom. 2010. *Market Prospects for the iPhone and Its Economic Implications*. June. http://www.digieco.co.kr/KTFront/report/report_strategy_view.action?board _seq=3846&board_seq=3846&board_id=strategy

Korea Times. 1999. "The Number of Mobile Phone Subscribers Hit 20 Million." August 30.

Korea Times. 2015. "Middle-Aged Actors Rule Mobile Ads." October 5. http://www.koreatimes.co.kr/www/news/nation/2015/10/116_187997.html. Accessed November 28, 2015.

Korea Trade Investment Promotion Agency. N.d. *Analysis of Korea's Mobile Communication Industry*. Seoul: KOTRA.

Larson, James. 1995. *The Telecommunications Revolution in Korea*. New York: Oxford University Press.

Larson, James, and Jaemin Park. 2014. "From Developmental to Network State: Government Restructuring and ICT-Led Innovation Korea." *Telecommunications Policy* 38: 344–59.

Lechner, Frank. 2009. *Globalization: The Making of World Society*. Malden, MA: Wiley-Blackwell.

Lee, Dong-Hoo. 2012. "In Bed with the iPhone: The iPhone and Hypersociality in Korea." In *Studying Mobile Media: Cultural Technologies, Mobile Communication, and the iPhone*, ed. Larissa Hjorth, Jean Burgess, and Ingrid Richardson, 63–81. London: Routledge.

Lee, Dong-Hoo. 2013. "Smartphones, Mobile Social Space, and New Sociality in Korea." *Mobile Media and Communication* 1 (3): 269–84.

Lee, Heejin, Robert M. O'Keefe, and Kyounglim Yun. 2003. "The Growth of Broadband and Electronic Commerce in South Korea: Contributing Factors." *Information Society* 19 (1): 81–93.

Lee, Heejin, and Sangjo Oh. 2008. "The Political Economy of Standards Setting by Newcomers: China's WAPI and South Korea's WIPI." *Telecommunications Policy* 32 (9–10): 662–71.

Lee, Hyungoh, and Sang-young Han. 2002. "The Evolution of the National Innovation System in the Korean Mobile Telecommunication Industry." *Communications and Strategies* 48: 161–86.

Lee, HyunJoo, Namsu Park, and Yongsuk Hwang. 2015. "A New Dimension of the Digital Divide: Exploring the Relationship between Broadband Connection, Smartphone Use and Communication Competence." *Telematics and Informatics* 32 (1): 45–56.

Lee, H. G. 2011. "Social and Cultural Meanings of Mobile Phone Adoption in Korea." In *The Korean Society for Communication Studies*, ed. Digital Media and Culture in Korea, 215–48. Seoul: Communication Books.

Lee, Jong Hyul, and Junghyun Kim. 2014. "Socio-demographic Gaps in Mobile Use, Causes, and Consequences: A Multi-group Analysis of the Mobile Divide Model." *Information, Communication and Society* 17 (8): 917–36.

Lee, Jung-ah. 2009. "LG Telecom Merger Wins Antitrust Approval." December 3. *Wall Street Journal*. http://online.wsj.com/news/articles/SB10001424052748704107104574573350250429942. Accessed October 5, 2013.

Lee, Jungah, and Jason Folkmanis. 2013. "Samsung Shifts Plants from China to Protect Margins." *Bloomberg News*. December 11. http://www.bloomberg.com/news/2013-12-11/samsung-shifts-plants-from-china-to-protect-margins.html. Accessed February 2, 2014.

Lee, Kwang Suk. 2008. "Globalization, Electronic Empire and the Virtual Geography of Korea's Information and Telecommunications Infrastructure." *International Communication Gazette* 70 (1): 3–20.

Lee, Kwang Suk. 2011. *IT Development in Korea: A Broadband Nirvana?* London: Routledge.

Lee, Min-jeong, and Jonathan Cheng. 2014. "In Samsung Country, iPhone 6 Fans Will

Have to Wait." September 11. http://blogs.wsj.com/digits/2014/09/11/in-samsung-country-iphone-6-fans-will-have-to-wait/. Accessed December 5, 2014.

Lee, Youkyung. 2010. "Mobile Big Bang Strikes S. Korea's Smartphone Market." *Yonhap News*. December 20.

Lemos, Andre. 2011. "Pervasive Computer Games and Processes of Spatialization: Informational Territories and Mobile Technologies." *Canadian Journal of Communication* 36: 277–94.

Levie, Aaron. 2013. "The Enterprise App Economy." June 8. http://techcrunch.com/2013/06/08/the-enterprise-app-economy/. Accessed November 5, 2014.

Lev-Ram, M. 2013. "Samsung's Road to Global Domination." *Fortune*. January 22. http://fortune.com/2013/01/22/samsungs-road-to-global-domination/

Lewis, Peter. 1993. "The Executive Computer; Bulky and Costly, but a Portable Succeeds." *New York Times*. October 17. http://www.nytimes.com/1993/10/17/business/the-executive-computer-bulky-and-costly-but-a-portable-succeeds.html

Lim, Sun Sun, and Gerald Goggin. 2014. "Mobile Communication in Asia: Issues and Imperatives." *Journal of Computer-Mediated Communication* 19: 663–66.

Ling, Richard. 2008. *New Tech, New Ties: How Mobile Communication Is Reshaping Social Cohesion*. Cambridge, MA: MIT Press.

Ling, Richard. 2012. *Taken for Greatness*. Cambridge, MA: MIT Press.

Ling, Richard, and Dag Svanes. 2011. "Browsers vs. Apps: The Role of Apps in the Mobile Internet." Paper presented to the conference "Internet and Society: Challenges, Transformation and Development," December.

Ling, Richard, and B. Yttri. 2002. "Hyper-coordination via Mobile Phones in Norway." In *Perpetual Contact: Mobile Communication, Private Talk, Public Performance*, ed. J. E. Katz and M. Aakhus, 139–69. Cambridge, MA: Cambridge University Press.

Linke, Christine. 2015. "Mobile Media and Communication in Everyday Life: Milestones and Challenges." *Mobile Media and Communication* 1 (1): 32–37.

Livingstone, Sonia, and E. Helsper. 2007. "Gradations in Digital Inclusion: Children, Young People and the Digital Divide." *New Media and Society* 9 (4): 671–96.

Mac, Ryan. 2014. "Mobile Master: KakaoTalk Creator Becomes One of South Korea's Richest Billionaires." *Forbes*. September 24. http://www.forbes.com/sites/ryanmac/2014/09/24/mobile-master-kakaotalk-creator-becomes-one-of-south-koreas-richest-billionaires/

Mac, Ryan. 2015. "How Kakao Talk's Billionaire Creator Ignited a Global Messaging War." *Forbes*. March 2. http://www.forbes.com/sites/ryanmac/2015/03/02/kakao-talk-billionaire-brian-kim-mobile-messaging-global-competition/

Mackenzie, Donald, and Judy Wajcman. 1999. *The Social Shaping of Technology: How the Refrigerator Got Its Hum*. Philadelphia: Open University Press.

MacMillan, Douglas. 2009. "Inside the App Economy." *Business Week*. October 22. http://www.businessweek.com/magazine/content/09_44/b4153044881892.htm. Accessed March 1, 2013.

Mahlich, Jorg, and Werner Pascha, eds. 2012. *Korean Science and Technology in an International Perspective*. Physica.

Mancinelli, Elisa. 2007. *E-Inclusion in the Information Society*. www.ittk.hu/netis/doc/ISCB_eng/10_Mancinelli_final.pdf. Accessed May 6, 2014.

Mandel, Michael. 2012. "Where the Jobs Are: The App Economy." *Technet*. http://www.

technet.org/wp-content/uploads/2012/02/TechNet-App-Economy-Jobs-Study.pdf. Accessed December 1, 2015.

Maney, Kevin. 1993. "Simon Says: Super-phone Is Giant Step." *USA Today*. November 3, 2B.

Manovich, Lev. 2013. *Software Takes Command*. London: Bloomsbury.

Mansell, Robin. 2002. "From Digital Divides to Digital Entitlements in Knowledge Societies." *Current Sociology* 50 (3): 407–26.

Marketing. 2013. "The Last-Mover Advantage." January 23.

Marketwire. 2012. "App Economy under Attack." August 20. http://www.marketwire.com/printer_friendly?id=1692285

Marvin, Bob. 2015. "How the Mobile App Economy Will Be Won." *PC Magazine*. October 2. http://www.pcmag.com/article2/0,2817,2492454,00.asp. Accessed December 1, 2015.

Marx, Leo, and Merritt Roe Smith. 1994. *Does Technology Drive History? The Dilemma of Technological Determinism*. Cambridge, MA: MIT Press.

Mascheroni, Giovanna, and Kjartan Olafsson. 2015. "The Mobile Internet: Access, Use, Opportunities and Divides among European Children." *New Media and Society* (online first). 1–23.

McCarty, Brad. 2011. "The History of the Smartphone." *TNW Blog*. December 6. http://thenextweb.com/mobile/2011/12/06/the-history-of-the-smartphone/#!xW2Cu. Accessed November 4, 2013.

McChesney, Robert. 2008. *The Political Economy of Media: Enduring Issues, Emerging Dilemmas*. New York: Monthly Review Press.

McChesney, Robert. 2013. *Digital Disconnect: How Capitalism Is Turning the Internet against Democracy*. New York: New Press.

McChesney, Robert, and Dan Schiller. 2003. "The Political Economy of International Communications: Foundations for the Emerging Global Debate over Media Ownership and Regulation." UNRISD Project on Information Technologies and Social Development. December 1. 1–33.

McKelvie, Alexander, and Richard Picard. 2008. "The Growth and Development of New and Young Media Firms." *Journal of Media Business Studies* 5 (1): 1–8.

McLelland, Mark. 2007. "Socio-cultural Aspects of Mobile Communication Technologies in Asia and the Pacific: A Discussion of the Recent Literature." *Continuum* 21 (2): 267–77.

McLuhan, Marshall. 1994. *Understanding Media: The Extensions of Man*. Cambridge, MA: MIT Press.

McVeigh, Brian. 2003. "Individualization, Individuality, Interiority, and the Internet: Japanese University Students and E-mail." In *Japanese Cybercultures*, ed. Nanette Gottlieb and Mark McLelland, 19–33. London: Routledge.

Melody, William. 2009. "Markets and Policies in New Knowledge Economies." In *The Oxford Handbook of Information and Communication Technologies*, ed. Robin Mansell, C. Avgerou, D. Quah, and R. Silverstone, 55–74. New York: Oxford University Press.

Mervyn, Kieran, Anoush Simon, and David Allen. 2014. "Digital Inclusion and Social Inclusion: A Tale of Two Cities." *Information, Communication and Society* 17 (9): 1086–1104.

Miller, Toby. 2006. "Gaming for Beginners." *Games and Culture* 1 (1): 5–12.

Miller, Toby, and M. Leger. 2001. "Runaway Production, Runaway Consumption, Runaway Citizenship: The New International Division of Cultural Labor." *Emergence* 11 (1): 89–105.

Millward, Steven. 2013. "In a Major Milestone, Korean-Made Kakao Talk Reaches 100 Million Users." *TechinAsia*. July 2. http://www.techinasia.com/kakaotalk-reaches-100-million-users/. Accessed June 3, 2016.

Mims, Christopher. 2016. "Why Microsoft Bought LinkedIn." *Wall Street Journal*. June 14. http://www.wsj.com/articles/microsoft-gains-link-to-a-network-1465922927. Accessed June 17, 2016.

Ministry of Culture and Tourism (Munhwa Kwan'gwang Bu). 2007. *2007 Taehan Min'guk keim paeksŏ* (*2007 White Paper on Korean Games*). Seoul: Munhwa Kwan'gwang Bu.

Ministry of Culture, Sports and Tourism (Munhwa Ch'eyuk Kwan'gwang Bu). 2012. *2011 k'ont'ench'ŭ sanŏp paeksŏ* (*2011 Contents Industry Whitepaper*). Seoul: Munhwa Ch'eyuk Kwan'gwang Bu.

Ministry of Culture, Sports and Tourism. 2013. *2013 Taehan Min'guk keim paeksŏ* (*2013 White Paper on Korean Games*). Seoul: Munhwa Ch'eyuk Kwan'gwang Bu.

Ministry of Culture, Sports and Tourism. 2014. *2014 Taehan Min'guk keim paeksŏ* (*2014 White Paper on Korean Games*). Seoul: Munhwa Ch'eyuk Kwan'gwang Bu.

Ministry of Culture, Sports and Tourism. 2015. *2015 Taehan Min'guk keim paeksŏ* (*2015 White Paper on Korean Games*). Seoul: Munhwa Ch'eyuk Kwan'gwang Bu.

Ministry of Finance and Economy of Korea (Chaejŏng Kyŏngje Pu). 2003. "Woegugin chikchŏp t'uja e taehan chose chiwŏn chedo ŭi sŏnggwa mit hyanghu unyong pangan (ch'oejong pogosŏ)" (Liberalization of FDI in Principle). http://mofe.go.kr/mofe/kor/fdi/html/e-2-1.htm. Accessed February 1, 2013.

Ministry of Gender, Equality and Family (Yŏsŏng Kajok Pu). 2013. *2013 ch'ŏngsonyŏn maech'e iyong silt'ae chosa* (*2013 Youth Media Use Report*). Seoul: Yŏsŏng Kajok Pu.

Ministry of Information and Communication (Chŏngbo T'ongsinbu). 2004. *The Road to 20,000 GDP/Capita*. Seoul: MIC.

Ministry of Information and Communication. 2008. *SK t'ellek'om ŭi hanaro t'ellek'om chusik ch'uidŭk inga simsa kyŏlgwa* (*The Result of Investigation for Approval of SKT's Acquisition of Hanaro Telecom*). February.

Ministry of Knowledge Economy. 2013. *IT Export Achieved Net Gains in 2012*. Seoul: MKE.

Ministry of Science, ICT and Future Planning (Mirae Ch'angjo Kwahak Pu). 2013. *Kukka chŏngbohwa e kwanhan yŏnch'a bogosŏ* (*2013 Annual Report of the Nation Information Status*). Seoul: Mirae Ch'angjo Kwahak Pu.

Ministry of Science, ICT and Future Planning. 2014a. *Yusuhan t'onggye (2013)* (*Status of Wireless and Wire Telephone Lines of 2013*). Seoul: Mirae Ch'angjo Kwahak Pu.

Ministry of Science, ICT and Future Planning. 2014b. *Future Mobile Telecommunications Industry Development Plan for the Creative Economy*. Press release. Seoul: Mirae Ch'angjo Kwahak Pu.

Ministry of Science, ICT and Future Planning. 2015a. *15-nyŏn 9-wŏl musŏn t'ongsin sŏbisŭ t'onggye hyŏnhwang* (*Status of Wireless Telephone Lines of September of 2015*). Seoul: Mirae Ch'angjo Kwahakpu.

Ministry of Science, ICT and Future Planning. 2015b. "2014 ICT Exports Surpassed $1738 million." Press release. January 9.

Ministry of Science, ICT and Future Planning. 2016. *16-nyŏn 3-wŏl musŏn t'ongsin sŏbisŭ t'onggye hyŏnhwang* (*Status of Wireless Telephone Lines of March of 2016*). Seoul: Mirae Ch'angjo Kwahakpu.

Molnar, Szilard. 2003. "The Explanation Frame of the Digital Divide." Proceedings of the IFIP summer school "Risks and Challenges of the Networked Society," 4–8. Karlstad University, August.

Moon, Ihlwan. 2008. "Why Korea Won't Bite Apple." *Businessweek*. June 13. http://www.businessweek.com/stories/2008-06-13/why-korea-wont-bite-the-applebusinessweek-business-news-stock-market-and-financial-advice. Accessed December 4, 2013.

Mosco, Vincent. 2014. *To the Cloud: Big Data in a Turbulent World*. Boulder, CO: Paradigm.

Munhwa Ilbo. 1999. "Hyudae chŏnhwa kaipcha ol ane 2200-man nŏmŭl tŭt" (Mobile Phone Subscribers Reached over 22 Million). September 1, 14.

Murphy, Dean. 2000. "Two Continents, Disconnected: Europeans Are Finding New York to Be a Backwoods for Cell Phones." *New York Times*. December 14, B1.

Nam Hye-hyŏn. 2014. "Kugŭl, Aep'ŭl, 3-cho wŏn tae kungnae aep sijang ssakssŭli" (Google and Apple Sweep $3 Billion Worth Domestic App Market). ZDNet Korea, ZDNet.co.kr. May 11. http://www.zdnet.co.kr/news/news_view.asp?artice_id=20140529133511. Accessed November 9, 2014.

Narayanaswamy, Shankar, Jianying Hu, and Ramanujan Kashi. 1998. "Using Data on Digital Cellular and PCS Voice Networks." *Bell Labs Technical Journal*. April–June, 58–75.

NAS Media. 2015. *Netizen Profile Report*. Seoul: NAS Media.

Nath, Asoke, and Mukherjee, Sneha. 2015. "Impact of Mobile Phone/Smartphone: A Pilot Study on Positive and Negative Effects." *International Journal of Advance Research in Computer Science and Management Studies* 3 (5): 294–302.

Natanson, Elad. 2015. "The State of Mobile and the App Economy in 2015." *Forbes*. May 26. http://www.forbes.com/sites/eladnatanson/2015/05/26/the-state-of-mobile-and-the-app-economy-in-2015/

Nielsen. 2012a. "America's New Mobile Majority: A Look at Smartphone Owners in the U.S." May 7. Available at http://blog.nielsen.com/nielsenwire/online_mobile/who-owns-smartphones-in-the-us/

Nielsen. 2012b. "Mobile App vs. Mobile Web." http://mobizen.pe.kr/1154

Nielsen. 2013. *The Mobile Consumer: The Global Snapshot*. New York: Nielsen.

Nielsen. 2014. "An Era of Growth: The Cross-Platform Report Q4 2013." March 5. http://www.nielsen.com/us/en/insights/reports/2014/an-era-of-growth-the-cross-platform-report.html. Accessed December 6, 2015.

Nielsen Asia. 2012. "Smartphone Ownership on the Rise in Asia Pacific, Whilst Advertisers Struggle to Engage with Consumers via Mobile Ads." June 20. http://jp.en.nielsen.com/site/documents/SPImr-jun12_FINAL.pdf. Press release.

Norris, Pippa. 2001. *Digital Divide: Civic Engagement, Information Poverty, and the Internet Worldwide*. New York: Cambridge University Press.

O'Malley, C. 1994. "Simonizing the PDA." *Byte*. December 20.

Oh, Myung, and James Larson. 2011. *Digital Development in Korea: Building an Information Society*. London: Routledge.

Ohmae, Konichi. 1995. "The End of the Nation State: The Rise of Regional Economy." *Foreign Affairs*. July–August.

Ok, Hye Ryoung. 2011. "New Media Practices in Korea." *International Journal of Communication* 5: 320–48.

Oliver, Christian, and Jung-a Song. 2009. "S. Korean Regulators Relent on iPhone Sales." *Financial Times*. September 23.

Organization for Economic Cooperation and Development. 2013a. *The App Economy*. December 17. Paris: OECD.

Organization for Economic Cooperation and Development. 2013b. *Communications Outlook 2013*. Paris: OECD.

Organization for Economic Cooperation and Development. 2014. *OECD Economic Surveys Korea*. Paris: OECD.

Orlando Sentinel. 2013. "IBM's Simon Personal Communicator Was Years Ahead of Its Time." March 25. http://articles.orlandosentinel.com/2013-03-25/business/os-cfb-talking-with-rich-guidotti-20130325_1_wireless-carriers-wireless-technology-business-person. Accessed November 26, 2013.

Pacey, Arnold. 1993. *The Culture of Technology*. Cambridge, MA: MIT Press.

Pak Tong-gyu and Kim Sŏng-gwan. 2014. "Sil teit'o sujip ŭl t'onghan sŭmat'ŭp'on iyongja ŭi aep'ŭllik'eisyŏn sayong sigan kwa iyong p'aet'ŏn punsŏk" (An Analysis of Smartphone Use Pattern through a Real Time Access). *Chugan kisul tonghyang 1649-ho* (*Weekly IT Trend 1649*). June 11, 1–19. Seoul: Chŏngbu t'ongsin sanŏp chinhŭngwŏn (National IT Industry Promotion Agency).

Pak Yŏng-ju. 2014. "K'ak'aot'ok sijang chŏmyuyul 92%" (Kakao Talk's Market Share 92%). *Nyusisŭ* (*Newsis*). September 23. http://www.newsis.com/ar_detail/view.html?ar_id=NISX20140923_0013187317&cID=10402&pID=10400. Accessed November 6, 2014.

Park, C. K. 2009. "iPhone's Debut in S. Korea Means Paradigm Shift: Experts." AFP. November 28.

Park, Eun-a. 2014. "Exploring the Multidimensionality of the Smartphone Divide: A New Aspect of the Digital Divide." Paper presented at the 36th Annual Pacific Telecommunications Conference. Honolulu, HI, January 19–22.

Park, Jong H. 2002. "The East Asian Model of Economic Development and Developing Countries." *Journal of Developing Societies* 18 (4): 330–53.

Park, Young Jin. 2015. "My Whole World's in My Palm! The Second-Level Divide of Teenagers' Mobile Use and Skill." *New Media and Society* 17 (6): 977–95.

Perez, Sarah. 2014. "Mobile App Usage Increases in 2014, as Mobile Web Surfing Declines." April 1. http://techcrunch.com/2014/04/01/mobile-app-usage-increases-in-2014-as-mobile-web-surfing-declines/. Accessed February 27, 2015.

Pieterse, Jan Nederveen. 2006. "Neoliberal Globalization and the Washington Consensus." In *International Development Governance*, ed. Ahmed S. Huque and Habib M. Zafarullah, 91–104. London: Routledge.

Pinch, Trevor, and Wiebe Bijker. 1984. "The Social Construction of Facts and Artefacts: Or How the Sociology of Science and the Sociology of Technology Might Benefit Each Other." *Social Studies of Science* 14 (3): 399–441.

Presidential Speeches (Taet'ongnyŏng yŏnsŏl). 2012. "Yi Myŏng-bak Taet'ongryŏng che 67-hoe kwangbukchŏl kyŏngch'uksa" (Address by President Lee Myung-bak on the 67th Anniversary of Liberation). Press release. September 16.

"Product of the Month." 1994. "BellSouth Cellular/IBM Release Simon PDA." *Telecommunications* 28 (1): 116.

Puspitasar, Liar, and Kenichi Ishii. 2015. "Digital Divides and Mobile Internet in Indonesia: Impact of Smartphones." *Telematics and Informatics* (online first): 1–12.

Qiu, Jack. 2009. *Working-Class Network Society: Communication Technology and the Information Have-Less in Urban China*. Cambridge, MA: MIT Press.

Qiu, Jack, and Yeran Kim. 2010. "Global Financial Crisis Recession and Progression? Notes on Media, Labor, and Youth from East Asia." *International Journal of Communication* 4: 630–48.

Quan-Haase, Anabel. 2013. *Technology and Society: Social Networks, Power, and Inequality*. New York: Oxford University Press.

Rabouin, Dion. 2015. "Huawei \$46 Billion in Sales Revenue Highlights Growing Boom of Chinese Companies in Smartphone Wars." *International Business Times*. January 3. http://www.ibtimes.com/huawei-46-billion-sales-revenue-highlights-growing-boom-chinese-companies-smartphone-1772682. Accessed February 27, 2015.

Rainie, Lee, and Berry Wellman. 2012. *Networked: The New Social Operating System*. Cambridge, MA: MIT Press.

Ramstad, Evan. 2009. "iPhone Tries to Crack Korea: Samsung Slashes Price of Its High-End Device Ahead of Apple's Debut." *Wall Street Journal*. November 27. http://online.wsj.com/news/articles/SB10001424052748703499404574559734131133944. Accessed January 30, 2014.

Rheingold, Howard. 2002. *Smart Mobs: The Next Social Revolution*. New York: Basic Books.

Richardson, Ingrid. 2011. "The Hybrid Ontology of Mobile Gaming." *Convergence* 17 (4): 419–30.

Richardson, Ingrid. 2012. "Touching the Screen: A Phenomenology of Mobile Gaming and the iPhone." In *Studying Mobile Media: Cultural Technologies, Mobile Communication, and the iPhone*, ed. Larissa Hjorth, Jean Burgess, and Ingrid Richardson, 133–53. London: Routledge.

Ritzer, George. 2011. *Globalization: The Essentials*. Malden, MA: Wiley-Blackwell.

Robertson, J. 1994. "Mitsubishi Puts Picture in the Hand." *Electronic Buyer's News*. December 12.

Rogers, Everett. 1983. *Diffusion of Innovation*. 3rd ed. New York: Free Press.

Rogers, Everett. 2003. *Diffusion of Innovation*. 5th ed. New York: Free Press.

Rose, Frank. 2001. "Pocket Monster: How DoCoMo's Wireless Internet Service Went from Fad to Phenom—and Turned Japan into the First Post-PC Nation." *Wired*. September (9). http://www.wired.com/2001/09/docomo/. Accessed December 2, 2013.

Rusli, Evelyn. 2013. "The Messaging Apps Taking on Facebook, Phone Giants." *Wall Street Journal*. March 27.

Russell, Jon. 2013. "Korean Messaging App KaKao Talk's Games Platform Grossed \$311 Million in HI 2013." *TNW*. July 16. http://thenextweb.com/asia/2013/07/16/korean-messaging-app-kakao-talks-games-platform-grossed-311-million-in-h1-2013/. Accessed February 2, 2014.

Sager, Ira. 2012. "Before iPhone and Android Came Simon, the First Smartphone." *Bloomberg Businessweek*. June 29.

Sahin, Ismail. 2006. "Detailed Review of Rogers's Diffusion of Innovations Theory and Educations Technology-Related Studies Based on Rogers' Theory." *Turkish Online Journal of Educational Technology* 5 (3): 14–23.

Sarwar, Muhammad, and Tariq R. Soomro. 2013. "Impact of Smartphone's on Society." *European Journal of Scientific Research* 98 (2): 216–26.

Schifferes, Steve. 2007. "Globalization Shakes the World." BBC News. January 21. http://news.bbc.co.uk/2/hi/business/6279679.stm. Accessed May 18 2013.

Schiller, Dan. 1982. "Business Users and the Telecommunications Network." *Journal of Communications* 32 (4): 84–96.

Schiller, Dan. 1999. *Digital Capitalism*. Cambridge, MA: MIT Press.

Schiller, Dan. 2007. *How to Think about Information*. Urbana: University of Illinois Press.

Schiller, Dan. 2012. *Digital Depression*. Urbana: University of Illinois Press.

Schiller, Dan. 2014. *Digital Depression: Information Technology and Economic Crisis*. Urbana: University of Illinois Press.

Schiller, Herbert. 1976. *Communication and Cultural Dominance*. New York: International Arts and Sciences Press.

Schrage, M. 1983. "MCI Weighs Sale of Smart Phones." *Washington Post*. August 5, E9.

Selwyn, Neil. 2004. "Reconsidering Political and Popular Understandings of the Digital Divide." *New Media and Society* 6 (4): 341–64.

Seoul Metro. 2014. "Daily Subway Riders." http://115.84.165.91/jsp/WWS00/outer_Seoul.jsp?stc_cd=69

Shade, Leslie Regan. 2007. "Feminizing the Mobile: Gender Scripting of Mobiles in North America." *Continuum* 21 (2): 179–89.

Shi, Yu. 2011. "iPhones in China: The Contradictory Stories of Media-ICT Globalization in the Era of Media Convergence and Corporate Synergy." *Journal of Communication Inquiry* 35 (2): 134–56.

Shin, Ji-hye. 2014. "Foreign Companies Eye Local Mobile Game Market." *Korea Herald*. July 20.

Siegal, Jacob. 2014. "Samsung Wants to Help You Manage All Your Devices with Just One App." *BGR*. January 6. http://bgr.com/2014/01/06/samsung-smart-home-connectivity-service/. Accessed May 9, 2014.

Silva, Adriana, and Larissa Hjorth. 2009. "Playful Urban Spaces: A Historical Approach to Mobile Games." *Simulation and Gaming* 40 (5): 602–25.

Silverstone, Roger, and Eric Hirsch, eds. 1992. *Consuming Technologies: Media and Communication in Domestic Spaces*. London: Routledge.

Sinckars, Pelle, and Patrick Vonderau, eds. 2012. *Moving Data: The iPhone and the Future of Media*. New York: Columbia University Press.

Sinclair, M., and M. Brown. 1983. "Phones on Threshold of Brave New World." *Washington Post*. January 2, A1.

Song, Jung-a. 2014. "S Korea's Kakao to Merge with Daum." *Financial Times*. May 28. http://www.ft.com/intl/cms/s/0/c97f2d1e-e483-11e3-a73a-00144feabdc0.html#axzz3EgzmxyWj

Sony. 2012. "Sony Completes Full Acquisition of Sony Ericsson." Press release. February 16.

Sŏul Kyŏngje (*Seoul Economic Daily*). 2012. "iOS kungnae chŏmyuyul 9.3% yŏktae ch'oejŏ" (iOS's Market Share Dropped to 9.3%). October 23.

Sparks, Colin. 2013. "What Is the Digital Divide and Why Is It Important." *Javnost: The Public* 20 (2): 27–46.

Statista. 2014. "Worldwide Market Share of Leading Search Engines from January 2010 to July 2014." http://www.statista.com/statistics/216573/worldwide-market-share-of-search-engines/

Stewart, Angus. 2000. "Social Inclusion: An Introduction." In *Social Inclusion: Possibilities and Tensions*, ed. Peter Askonas and Aagus Stewart, 1–16. New York: Palgrave.

Stiglitz, Joseph. 2003. *Globalization and Its Discontents*. New York: Norton.

Stockholm Smartphone. 2014. "Stockholm Smartphone." http://www.stockholmsmartphone.org/. Accessed September 6, 2014.

Strover, Sharon. 2014. "The US Digital Divide: A Call for a New Philosophy." *Critical Studies in Media Communication* 31 (22): 114–22.

Summerfield, Jason. 2014. "Mobile Website vs. Mobile App (Application): Which Is Best for Your Organization?" Human Service Solutions. http://hswsolutions.com/services/mobile-web-development/mobile-website-vs-apps/. Accessed September 6, 2014.

Sung, Tae Kyung. 2009. "Technology Transfer in the IT industry: A Korean Perspective." *Technological Forecasting and Social Change* 76: 700–708.

Tai, Zixue, and Haifang Zeng. 2011. "Mobile Games in China: Formation, Ferment, and Future." In *Global Media Convergence and Cultural Transformation: Emerging Social Patterns and Characteristics*, ed. Dal Yong Jin, 270–95. New York: Information Science Reference.

Tapscott, Don. 1998. *Growing Up Digital: The Rise of the Net Generation*. New York: McGraw Hill.

Thussu, Daya. 2006. *International Communication: Continuity and Change*. London: Arnold.

Thussu, Daya. 2007. "Mapping Global Media Flow and Contra-Flow." In *Media on the Move*, ed. Daya Thussu, 11–32. London: Routledge.

Tomlinson, John. 2000. *Globalization and Culture*. Cambridge: Polity Press.

TrendForce. 2015. "Top 10 Smartphone Makers." Press release. January 20.

Tsatsou, Panayiota. 2011. "Digital Divides Revisited: What Is New about Divides and Their Research?" *Media, Culture and Society* 33 (2): 317–31.

U.S. Census Bureau. 2012. *Statistical Abstract of the U.S.* Washington, DC: U.S. Department of Commerce. http://www.census.gov/compendia/statab/cats/information_communications.html. Accessed May 6, 2014.

U.S. Department of Commerce. 1975. *Historical Statistics of the United States: Colonial Times to 1970*. Washington, DC: U.S. Department of Commerce.

U.S. Department of Commerce. 1999. *National Trade Data Bank*. September 3.

Usher, Willman. 2014. "61% of American Mobile Gamers Are Female." *Cinemablend*. http://www.cinemablend.com/games/61-American-Mobile-Gamers-Female-64391.html

Vanderhoef, John. 2013. "Casual Threats: The Feminization of Casual Video Games." *Journal of Gender, New Media and Technology* 2. http://adanewmedia.org/2013/06/issue2-vanderhoef/. Accessed December 5, 2013.

van Dijk, Jan A. G. M. 2005. *The Deepening Digital Divide: Inequality in the Information Society*. London: Sage.

van Dijk, Jan A. G. M. 2006. "Digital Divide Research, Achievements and Shortcomings." *Poetics* 34: 221–35.

van Dijk, Jan A. G. M. 2012a. *The Network Society*. London: Sage.

van Dijk, Jan A. G. M. 2012b. "The Evolution of the Digital Divide: The Digital Divide Turns to Inequality of Skills and Usage." *Digital Enlightenment Yearbook 2012*. 57–75.

van Dijk, Jan A. G. M. 2012c. "Facebook as a Tool for Producing Sociality and Connectivity." *Television and New Media* 13 (2): 160–76.

Verdegem, Pieter. 2011. "Social Media for Digital and Social Inclusion: Challenges for Information Society 2.0 Research & Policies." *Triple-C* 9 (1): 28–38.

Verkasalo, Hannu, Carolina López-Nicolás, Francisco Molina-Castillo, and Harry Bouwman. 2010. "Analysis of Users and Non-users of Smartphone Applications." *Telematics and Informatics* 27: 242–55.

Volti, Rudi. 2008. *Society and Technological Change*, 6th ed. New York: Worth Publishers.

Voskoglou, Christina. 2013. "Seizing the App Economy." DevelopersEconomics.com. http://www.developereconomics.com/report/sizing-the-app-economy/. Accessed June 7, 2016.

Wade, Robert. 2004. *Governing the Market: Economic Theory and the Role of Government in East Asian Industrialization*. 2nd ed. Princeton: Princeton University Press.

Waisbord, Silvio. 2004. "McTV: Understanding the Global Popularity of Television Formats." *Television and New Media* 5 (4): 359–83.

Waisbord, Silvio, and Claudia Mellado. 2014. "De-westernizing Communication Studies: A Reassessment." *Communication Theory* 24 (4): 361–72.

Ward, Andrew. 2004. "Where High-Speed Access Is Going Mainstream. The Korean Experience: Government Policy, High Levels of Urbanization." *Financial Times*. June, 4.

Warman, Peter. 2012. "Mid-core Gaming: Defining, Sizing and Seizing the Opportunity." *Newzoo*. April 18. http://www.newzoo.com/press-releases/mid-core-gaming-defining-sizing-and-seizing-the-opportunity/#hDFxr7Gs2HrPglKC.99. Accessed May 4, 2013.

Warmerdam, Marcel. 2014. "The App Economy: A Zero Sum Game." *Metis Files*. February 20. http://www.themetisfiles.com/2014/02/the-app-economy-a-zero-sum-game/. Accessed June 3, 2014.

Warschauer, Mark. 2002. "Reconceptualizing the Digital Divide." *First Monday* 7. http://firstmonday.org/article/view/967/888. Accessed May 3, 2013.

Warschauer, Mark. 2003. "Dissecting the Digital Divide: A Case Study in Egypt." *Information Society* 19 (4): 297–304.

Warschauer, Mark. 2004. *Technology and Social Inclusion: Rethinking the Digital Divide*. Cambridge, MA: MIT Press.

Watkins, Jerry, Larissa Hjorth, and Ilpo Koskinen. 2012. "Wising Up: Revising Mobile Media in an Age of Smartphones." *Continuum* 26 (5): 665–68.

WeMade. 2012. "K'aendip'ang sogae" (Introduce Candipang). http://social.wemade.com/game/game_info.asp?GmCode=7. Accessed May 3, 2013.

West, Joel, and Michael Mace. 2010. "Browsing as the Killer App: Explaining the Rapid Success of Apple's iPhone." *Telecommunications Policy* 34 (5–6): 270–86.

White, Andrew. 2014. *Digital Media and Society: Transforming Economics, Politics and Social Practices*. New York: Palgrave.

Williams, Raymond. 2003. *Television: Technology and Cultural Form*. 3rd ed. London: Routledge.

Wilson, Jason, Chris Chester, Larissa Hjorth, and Ingrid Richardson. 2011. "Distractedly Engaged: Mobile Gaming and Convergent Mobile Media." *Convergence* 17 (4): 351–55.

Wingfield, Nick. 2012. "Apple's Job Creation Data Spurs an Economic Debate." *New York Times*. March 4. http://www.nytimes.com/2012/03/05/technology/apple-study-on-job-creation-spurs-an-economic-debate.html?_r=0. Accessed May 6, 2013.

Winner, Langdon. 1993. "Upon Opening the Black Box and Finding It Empty: Social Constructivism and the Philosophy of Technology." *Science, Technology and Human Values* 18: 362–78.

Winseck, Dwayne. 2011. "The Political Economies of Media and the Transformation of the Global Media Industries." In *The Political Economies of Media*, ed. Dwayne Winseck and Dal Yong Jin, 3–48. London: Bloomsbury Academic.

Winston, Brian. 1998. *Media Technology and Society: A history. From the telegraph to the Internet*. London: Routledge.

Wireless. 2015. "SK Telecom Commercialises World's First Tri-Band LTE-A Service." January 5. http://www.wireless-mag.com/News/31845/sk-telecom-commercialises-world%E2%80%99s-first-tri-band-lte-a-service.aspx. Accessed June 7, 2016.

Wolf, Mark, ed. 2002. *The Medium of the Video Game*. Austin: University of Texas Press.

Yang, Sang Jin. 2002. "Mobile Phone Users Number 30 million, but Carriers Are Mired in Mudslinging." *Korea Herald*. April 5.

Yang, Sang Jin. 2003. "Convergence Key to Future Mobile Services." *Korea Herald*. September 25.

Yi Ch'ang-ho, Sŏng Yun-suk, Chŏng Nag-wŏn. 2012. *Chŏngsonyŏn ŭi sosyŏl midiŏ iyong silt'ae yŏngu* (*A Study on the SNS Use of Young People*). Seoul: Han'guk Chŏngsonyŏn Chŏngch'aek Yŏnguwŏn (National Youth Policy Institute).

Yi Chin-myŏng. 2014. "Samsŏng Chŏnja naenyŏn IoT ro sŭngbu gŏnda" (Samsung Electronics Selects IoT as the Next Driving Engine). *Maeil Kyŏngjae*. December 8. http://news.mk.co.kr/newsRead.php?year=2014&no=1506235. Accessed January 23, 2015.

Yi Hyŏng-gyŏng. 2014. "Han'guk intŏnet t'aedong kwa sŏngjang ŭi yŏksa" (The Birth and Growth of Korea's Internet). *Saiŏnsŭ taimjŭ* (*Science Times*). December 1. http://www.sciencetimes.co.kr/?news. Accessed January 26, 2015.

Yi Hyŏn-mi and Yi Chae-dong. 2012. "Sŭmat'ŭp'on kaein chŏngbo poho pisang" (Urgency in Protecting Personal Information on Smartphones). *Munhwa Ilbo*. October 8, 10.

Yi Tae-ho. 2015. "Net mabŭl, kaebalbi 100-ŏk wŏn tuip han taejak 'Idea" chŏt konggae" (Netmarble Opens Its Mobile Game IDEA). *Tijit'ŏl teilli*. May 20. http://www.ddaily.co.kr/news/article.html?no=130564

Yonhap News. 2010. "Mobile Big Bang Hits Korean Market." December 20.

Yonhap News (*Yŏnhap Nyusŭ*). 2012. "'K'at'ok ŏpsŭm motsara' . . . haru p'yŏnggyun 43-pun sayong" (No Life without Kakao Talk . . . the Average Daily Use of 43 Minutes). December 23. http://www.yonhapnews.co.kr/economy/2012/12/22/0303000000AKR20121222039800017.HTML. Accessed March 1, 2013.

Yonhap News. 2013. "Samsung Tops China's Smartphone Market in 2012." March 10. http://english.yonhapnews.co.kr/techscience/2013/03/10/34/0601000000AEN20130310001800320F.HTML. Accessed August 5, 2014.

Yonhap News. 2014a. "Mobile Carriers Release iPhone 6 in S. Korea." October 31. http://english.yonhapnews.co.kr/full/2014/10/31/33/1200000000AEN20141031004100320F.html. Accessed December 2, 2014.

Yonhap News. 2014b. "Android Rules S. Korean Market in 2013: Data." January 21. http://english.yonhapnews.co.kr/business/2014/01/21/30/0501000000AEN20140121001000320F.html. Accessed March 1, 2015.

Yonhap News (*Yŏnhap Nyusŭ*). 2014c. "Tosi kagu pindu kyŏkch'a kalsurok simhae chinda" (Income Inequality of Urban Households Is Widening). June 10. http://www.yonhapnews.co.kr/economy/2014/06/09/0301000000AKR20140609163900008.HTML. Accessed January 5, 2015.

Yoon, Chang Ho. 1999. "Liberalization Policy, Industry Structure and Productivity Changes in Korea's Telecommunication Industry." *Telecommunications Policy* 23: 289–306.

Yoon, Kyong. 2003. "Retraditionalizing the Mobile: Young People's Sociality and Mobile Phone in Seoul, South Korea." *European Journal of Cultural Studies* 6 (3): 327–43.

Yoon, Kyong. 2006. "The Making of Neo-Confucian Cyberkids: Representations of Young Mobile Phone Users in South Korea." *New Media and Society* 8 (5): 753–71.

Yoon, Sung-won. 2014a. "Blade Becomes First Mobile Title to Win Korea's Top Game Award." *Korea Times*. November 21. http://www.koreatimes.co.kr/www/news/tech/2014/11/134_168529.html

Yoon, Sung-won. 2014b. "LG UPlus Criticizes SKT, Samsung for Unfairness." *Korea Times*. June 23. http://www.koreatimes.co.kr/www/news/nation/2014/06/129_159665.html

Yoon, Sung-won. 2015a. "NCSOFT, Netmarble Tie Up against Nexon." *Korea Times*. February 17. http://www.koreatimes.co.kr/www/news/tech/2015/02/133_173838.html. Accessed May 2, 2015.

Yoon, Sung-won. 2015b. "Software Projects to Receive 100 Tril. in Loans." *Korea Times*. January 15. http://koreatimes.co.kr/www/news/tech/2015/01/129_171771.html

You, Soh Jung. 2003. "Lineage Creates Second Cyberworld Living Space." *Korea Herald*. October 8, 40.

Young, W. R. 1979. "Advanced Mobile Phone Service: Introduction, Background and Objective." *Bell System Technical Journal* 58 (1): 1–14.

Yu, K. H. 1998. "MIC Unveils Measures to Attract Foreign Direct Investment in Domestic IT Industry." *Korea Herald*. February 10, 26.

Yuan, Elaine J. 2012. "From 'Perpetual Contact' to Contextualized Mobility: Mobile Phones for Social Relations in Chinese Society." *Journal of International and Intercultural Communication* 5 (3): 208–25.

Zheng, Pei, and Lionel Ni. 2006. *Smart Phone and Next Generation Mobile Computing*. Amsterdam: Elsevier.

Zittrain, Jonathan. 2009. "Law and Technology: The End of the Generative Internet." *Communications of the ACM* 54 (1): 18–20.

Index

Note: Page numbers in *italics* refer to figures or tables.